The Chesapeake in F

THE CHESAPEAKE IN FOCUS

Transforming the Natural World

Tom Pelton

JOHNS HOPKINS UNIVERSITY PRESS
Baltimore

© 2018 Johns Hopkins University Press
All rights reserved. Published 2018
Printed in the United States of America on acid-free paper
9 8 7 6 5 4 3 2 1

Johns Hopkins University Press
2715 North Charles Street
Baltimore, Maryland 21218-4363
www.press.jhu.edu

Library of Congress Cataloging-in-Publication Data

Names: Pelton, Tom, 1967– author.
Title: The Chesapeake in focus : transforming the natural world /
  Tom Pelton.
Description: Baltimore : Johns Hopkins University Press, [2018] |
  Includes bibliographical references and index.
Identifiers: LCCN 2017025725| ISBN 9781421424750 (pbk. : acid-free
  paper) | ISBN 9781421424767 (electronic) | ISBN 1421424754
  (pbk. : acid-free paper) | ISBN 1421424762 (electronic)
Subjects: LCSH: Chesapeake Bay (Md. and Va.)—Environmental
  conditions. | Water pollution—Chesapeake Bay (Md. and Va.)
Classification: LCC GE155.C48 P45 2018 | DDC 577.7/270916347—
  dc23
LC record available at https://lccn.loc.gov/2017025725

A catalog record for this book is available from the British Library.

All photographs by the author. Map on page 2 is by Bill Nelson.

*Special discounts are available for bulk purchases of this book. For
more information, please contact Special Sales at 410-516-6936 or
specialsales@press.jhu.edu.*

Johns Hopkins University Press uses environmentally friendly book
materials, including recycled text paper that is composed of at least
30 percent post-consumer waste, whenever possible.

To my wife Liz, for everything

Contents

Acknowledgments

This book reflects the opinions and conclusions I've drawn from interviews and research I've performed as a writer and radio host covering the Chesapeake Bay for two decades. Some of the chapters expand on subjects I discussed on my weekly public radio program, *The Environment in Focus*, or wrote about in the *Baltimore Sun* or other publications. The radio show, which celebrated its tenth anniversary in 2017, is broadcast on NPR affiliate WYPR 88.1 FM in Baltimore and five other stations. Many thanks to WYPR for producing *The Environment in Focus* and for supporting my work, and to the Abell Foundation for sponsoring the radio program.

The views expressed in this book are entirely my own and do not necessarily reflect those of WYPR, the *Baltimore Sun*, the Chesapeake Bay Foundation, the Environmental Integrity Project, or any other organizations for which I have worked. I was a reporter at the *Baltimore Sun* from 1997 to 2008. Then, when the *Sun*'s parent company went into bankruptcy, the Chesapeake Bay Foundation (CBF) recruited me to serve as the organization's senior writer and investigative reporter, where I worked from 2008 to 2014. In May 2014, I decided to leave CBF to take a job as director of communications for the Environmental Integrity Project, a national nonprofit, nonpartisan organization based in Washington, DC, which specializes in environmental investigations and litigation. During much of this time, I also—from 2007 to the present—hosted the *Environment in Focus* public radio program and podcast. Although funded by a variety of philanthropic organizations over the years, including the Nature Conservancy and now the Abell Foundation, the radio program and podcast are owned by me and have always been independent of my work for other organizations.

The Chesapeake in Focus

Chesapeake Bay

Introduction

THE WAVES THREATENING CHESAPEAKE COUNTRY

I WAS IN AN AIRBOAT, with an engine as large as an airplane propeller, pounding across the Blackwater National Wildlife Refuge on Maryland's Eastern Shore. Islands of spartina grass rocketed past in waters that reflected sun like hammered silver. The captain, Bill Giese, an officer with the US Fish and Wildlife Service, jumped the boat over a muddy peninsula and then banked left, launching spray into the air. He slowed and then stopped. He pointed past a thicket of dead pine trees reaching their skeletal hands up out of the water to where a river flushed into the wetlands.

It was up that river, Giese explained, that 6,000 homes and a billion-dollar golf resort would soon be built around the entrance to the Blackwater National Wildlife Refuge, a 28,000-acre nature preserve that is sometimes called the "Everglades of the North." "I'm not opposed to development, but this would be just a huge explosion of growth all at once," Giese said.[1] "I'm worried that we're going to destroy the very thing people come here for." As he spoke, a great blue heron flapped overhead like a pterodactyl in a prehistoric swamp and a cloud of green-winged teal swirled into the sky from an island of reeds.

The development project was a surprise to me. Giese had taken me out on the water to show me something else: the damage that climate change was causing to the wildlife preserve. Rising sea levels caused by global warming were flooding more than 300 acres of grasses every year in the Blackwater River watershed, transforming marshes into open water. The rising seas were also driving salt water farther inland, killing pine trees and making the soil of farms too salty for crops.

But the rising seas were only one wave threatening the heart of Chesapeake country. The other deluge came from human population growth and suburban sprawl. About 5 miles north of Blackwater, along Egypt Road and Route 16 in Dorchester County, developers were planning three different projects that hijacked the word "Blackwater" into their advertising to lure home buyers into what had been an unspoiled area. The biggest was the proposed 3,200-home Blackwater Resort Community, with a golf course, tennis courts, pool, shops, parking lots, and a network of new roads to be built atop farm fields and wetlands. Nearby, bulldozers were already busy clearing land for the 793-home Blackwater Crossing subdivision and the 185-home Blackwater Landing. Together, all three would add about 7,000 new homes in a newly annexed strip of land—skinny and crooked as a gerrymander, stretching from the city of Cambridge far out into rural Dorchester County—that would make Cambridge's population of 12,326 more than double, nearly overnight. Because it would technically (although not really) be inside the boundaries of the historic but struggling city, the developers were touting it as "smart growth"—the concentration of new housing in existing communities. But in reality, it was the opposite: a magnet that would draw people out of the economically troubled town. The parking lots and driveways of the new subdivisions would also funnel hot rainwater runoff laced with oil, antifreeze, lawn fertilizer, and other pollutants into the Little Blackwater River, which dumps directly into the wildlife refuge.

It was a staggering assault on the last true wildlands in the Mid-Atlantic, and the construction threatened to contaminate one of the most productive breeding grounds for fish and birds in the Chesapeake region. Giese was the son and grandson of farmers who had hunted and plowed the lands of Dorchester County for decades. Ever since he was a boy, growing up next to the wildlife preserve, he had been canoeing and exploring the mosquito-infested paradise. For his entire adulthood, his job had been to care for the land and protect the more than 250 species of birds that migrated through or nested in the sanctuary, including tundra swans, golden eagles, swallows, osprey, and snowy egrets. He was determined to defend it all. "There's a lot of development coming, a lot of asphalt and rooftops," Giese said. "The system just can't take what's being put into it."

I knew what Giese was pointing to, up the river. He was pointing into the future, not only for the Blackwater but for the entire Chesapeake region—and for America. Reckless real estate speculation and poorly planned developments were destroying unique landscapes across the United States, leaving in their wake identical subdivisions, historic cities drained of people, and, eventually, bankruptcies and the worst economic collapse since the Great Depression.

At the time, in 2005, I was working for the *Baltimore Sun*, covering the environmental beat, and I'd soon launch a weekly public radio program called *The Environment in Focus*. But I had witnessed the same story unfold many times in the years before that, when I was writing for the *Boston Globe, Chicago Tribune, New Haven Register*, and *Charlottesville* (Virginia) *Daily Progress*. As I moved around the country learning the news business, I usually got my foot in the door by covering local government. I spent nights at county zoning hearings and days driving out to talk to neighbors of new development projects. The pattern was always the same: fields and forests, and unique and beautiful towns, destroyed for cookie-cutter subdivisions and chain stores.

This sterilization of the landscape was all the more offensive because it was now defacing the Chesapeake Bay, one of the world's ecological masterpieces. The Chesapeake is well known as the nation's largest estuary, a one-time epicenter of the oyster and crabbing industries, and a birthplace of American history. At its southern end, the bay boasts Jamestown, the first lasting English colony in North America and the seedbed of slavery, democracy, the tobacco industry, and the extirpation of Native Americans from much of the continent. Along the bay's western flank are Richmond and Washington, DC, rival capitals of the Civil War. Near the bay's head is Baltimore, the young republic's place of victory in the War of 1812 and the source of our nation's first railroad, the Baltimore & Ohio. To the bay's east is Delmarva, home of the great abolitionists Frederick Douglass and Harriet Tubman as well as Arthur Perdue, inventor of factory-style farming methods that have since spread around the world, "chickenizing" the global agricultural system.[2]

The Chesapeake is 200 miles long, stretching from the mouth of the Susquehanna River in northern Maryland to its opening on the Atlantic Ocean in Virginia. But the bay's watershed is much larger—a 64,000-square-mile basin that stretches across six states, from New York to Virginia, as well as the District of Columbia. The bay is a crossroads of life and culture, mixing the North with the South, the ocean's salt water with freshwater. When you look at the bay from an airplane, it is obvious why it is such an amazingly productive breeding ground for crabs, fish, and birds. The estuary's 11,684 miles of shoreline—more than the US west coast—are folded into countless nooks and crannies like the chambers of great green lungs. About 3,600 species of plants and animals shelter in these inlets, including 348 varieties of finfish and 173 kinds of shellfish. And about a million migratory waterfowl—including tundra swans, Canada geese, and canvasback ducks—stop over every fall and spring to feed, rest, and remind people why the Chesapeake became famous as a haven for hunters and bird-watchers.

Into this body every day pours an average of 50 billion gallons of water from more than 100,000 streams, creeks, and rivers.[3] About 18 million people live in the watershed, and about 150,000 more move in each year. Wastewater is dumped into the bay from 3,582 municipal sewage and water treatment plants and 1,679 industrial sites. Every time it rains, water flushes over 1.3 million acres of roofs, driveways, parking lots, and roads in the watershed, as well as from 87,000 farms with 6.5 million acres of crops.[4] This runoff picks up soil, fertilizer, oil, copper dust from brake pads, and other pollutants, and the toxic cocktail pours into an estuary that is extraordinarily shallow. Although the deepest parts of the bay are about 174 feet deep, its average depth is 21 feet, and about one-quarter of the bay's 4,480 square miles (including the estuary's tidal tributaries) are 6 feet or shallower, meaning that a person of my height and taste for slogging can wade right across them. The lack of depth (and therefore dilution) makes the bay unusually vulnerable to pollution. In fact, the land-to-water ratio in the Chesapeake watershed is the highest of any coastal water body in the world, 14:1—an unfortunate distinction, in terms of the bay's ability to absorb insults from the land.

The bay's inherent fragility has been knocked to the breaking point, and the blows started long ago. The English explorer Captain John Smith described the bay as a near paradise when he arrived in 1607. He wrote, "Heaven and Earth never agreed to frame a better place for man's habitation. Here are mountains, hills, plains, valleys, rivers, and brooks running into a fair bay, compassed but for the mouth with a fruitful and delightsome land." Fish were so numerous, colonist John Rolfe reported after his arrival in 1610, that a pair of men could wade into a stream and easily slay 40 sturgeon with axes, with each fish weighing 100 pounds (producing 4,000 pounds of food). Oysters were so plentiful, ships wrecked on the reefs. Over the next century and more, European landowners and their African slaves slashed, burned, and cleared forests to grow North America's first cash crop: tobacco. These English farmers often stole the corn-fields of the Native Americans to convert the land to plantations, worked by their slaves. The deforestation of the Chesapeake to carve out the new "tobacco coast" stripped away many of the bay's original water pollution control filters. This allowed rain to flush dirt and silt into once-rocky tributaries, smothering the hard-bottomed habitat that sturgeon, oysters, and other creatures need to survive. Tobacco was a highly lucrative crop, because it was addictive. But it was also a difficult plant that exhausted the soil and required farmers to keep moving to new fields and clearing more land, leaving behind an impoverished landscape.[5] The environmental damage from tobacco farming grew with the expanding human population in the nineteenth and twentieth centuries. Watermen overfished oysters, American shad, and sturgeon, decimating many of the bay's most important species. Meanwhile, cities such as Baltimore, Washington, DC, and Richmond employed their rivers as sewers for the dumping of raw human waste. By the 1960s and 1970s, the Chesapeake Bay was on its deathbed. Toxic algal blooms blanketed the Potomac and James Rivers, and foul odors wafted from Baltimore's harbor.

In an attempt to bring the bay back to life, the governments of Maryland, Virginia, Pennsylvania, and the District of Columbia signed cleanup agreements with the US Environmental Protection Agency (EPA) in 1983, 1987, and 2000 that were designed to restore the bay to ecological health by deadlines of 2000 and then 2010. Overall, there was a slight decrease in the level of nitrogen pollution in the bay—a nutrient that feeds algal blooms and low-oxygen "dead zones"—between the 1980s and about the year 2002, in large part because of upgrades to sewage treatment plants.[6] But then the bay's health stagnated or even slid backward from 2002 to 2010, for reasons that may have included suburban sprawl, unfavorable amounts of rainfall that flushed in more runoff pollution, and federal incentives for planting more corn, which demands tons of nitrogen fertilizer.

Then something encouraging happened around 2010: a hint of a possible turnaround for the bay after years in the doldrums. Most of the ecologically important trends started a slight and delicate dance upward, with less nitrogen pollution, and more dissolved oxygen, oysters, and bottom-dwelling clams and worms. Most impressively, underwater grasses in the bay—a key indicator and habitat for life beneath the waves—spread from 79,675 acres in 2010 to 97,433 acres in 2016, the largest expanse of aquatic vegetation since monitoring began more than three decades ago.[7] The bay's overall health improved from a score of 40 (out of 100) in 2011 to a 54 in 2016, according to the University of Maryland Center for Environmental Science's (UMCES) annual report cards based on water quality monitoring.[8] This modest and tentative improvement could have been influenced by a decline in rainfall and therefore runoff pollution (which, of course, could easily reverse itself). More significantly, the fragile progress could be crushed by bad policy decisions—a real threat, given the growing strength of anti-government, anti-regulatory political forces.

But still, any progress is worth celebrating and understanding. This is important because we need to defend what is working to improve the bay. And we need to move away from what is not working. The bay cleanup effort of the past three decades has been—if anything—a murky river of hype as well as hope, of broken promises as much as environmental leadership.

Returning to the scene on the Blackwater National Wildlife Refuge, Bill Giese eventually proved himself an environmental hero, winning an improbable war to stop the developers pushing the billion-dollar Blackwater Resort. By working over many months with neighbors, journalists, environmentalists, and others, Giese and allies convinced Governor Robert Ehrlich to buy and preserve most of the threatened land. It was a rare and spectacular victory—an example of what determined people can achieve, against all odds, to protect our natural world.

Who are the people really making a difference with the bay, and what forces are they up against? What are the policies driving the changes in the Chesapeake? And what impact are our policies having on life in the bay? These are all subjects I tackle in the following chapters, after first laying out the unique histories of, and challenges faced by, the rivers that flow into the bay like veins into a heart.

Now, hang on as I guide you on a boat ride through the swamplands.

The Waters

I N THIS FIRST SET of chapters, I'll take you on a tour of bay tributaries, leading you from north to south and past to present. We'll start with the Susquehanna River in Pennsylvania, the father of the bay and the wellspring of half its water and most of its problems.

Then we'll cast a wide net into the Gunpowder River in Baltimore County, to see what population growth is spawning beneath the waves.

Next, we'll motor down the Corsica River on Maryland's Eastern Shore to fish for clear lessons in a murky water quality experiment.

Then we explore the Patuxent River on the Western Shore, spending the day with an activist who grew up on the river and nearly sacrificed his life to it.

On the Potomac River, I'll swim among black blossoms of algae and swallow the fact that political polarization is the bay's worst pollutant.

As night falls over the James River in Virginia, I'll hear from the spirits of Jamestown about the dark heart of American history and the resurrection of a waterway.

And finally, I'll take a trip to the southern Chesapeake Bay to camp on some of its most pristine beaches and hear from a veteran naturalist about what really needs to be done to restore the waterway to health.

SUSQUEHANNA RIVER

*A Winter's Journey
to the Father of the Bay*

I T WAS A BLACK and icy morning as we dragged our kayaks down a rocky slope, littered with fast-food wrappers and a broken couch, and launched into the Susquehanna River. We slipped into darkness on the nearly mile-wide waterway that is the biggest source of freshwater in the Chesapeake Bay. About 3 miles southeast of Harrisburg, Pennsylvania, the temperature of the air and water was in the 30s—cold enough to kill a kayaker who had fallen in 10 days earlier.

My companions were Juan Veruete and Jeff Little, veteran fishing guides who had advised me how to dress to protect myself from hypothermia.[1] We wore waterproof dry suits that felt as stiff as spacesuits, as well as waterproof gloves and three layers of fleece.

Out on the river, the stars faded as the sky brightened to scarlet in the east. Winds whipped up small waves, which reflected the light as the sun rose over a line of leafless trees on the riverbank. Boulders jutted from the shallows.

For nearly 10 hours, we paddled on the Susquehanna, exposed in the biting wind, atop flat plastic floats the size of surfboards. I cast my rod with my right hand as I maneuvered my paddle with my left. It was a tricky game of fighting to stay pointed into the wind. Every now and then, a gust would blast my boat sideways, and I'd be blown away from my companions. To catch up with them again, I'd have to jam my fishing rod beneath a strap on my kayak and frantically dig at the water with my paddle, clawing my way back into control.

I thought, why do people fish—let alone go outside—in such conditions? Veruete said he heads out onto the Susquehanna in subfreezing temperatures all the time because he's been a fishing addict since he was five years old. He just can't keep himself off the river, even in potentially deadly conditions.

"It's almost like a drug, it's like an adrenaline junky kind of thing, you know?" Veruete said as he tugged his wool cap tighter. "You get out on the river, throw some baits, and you catch huge smallmouth bass. It's good for you."

The Susquehanna is the longest river on the East Coast, rambling 444 miles from Cooperstown in Upstate New York, through Pennsylvania dairy country, to Havre de Grace, Maryland, draining almost half of the farmland in the 64,000-square-mile Chesapeake Bay watershed.[2] The Susquehanna forms the northern end of the Chesapeake and provides about half of the freshwater for the estuary, about 19 million gallons per minute.

The river is one of the oldest in the world and predates the Chesapeake Bay by hundreds of millions of years. The Susquehanna's waters carved the path for the estuary, which is only about 10,000 years old. In that sense, the Chesapeake Bay—although the largest estuary in the United States—is but a baby of its craggy and ancient father.

Because it is so rocky, the Susquehanna is inhospitable to commercial boat traffic. To overcome this problem, during the early 1800s, developers built canals along the river's southern stretch. But these canals were soon abandoned for faster railroad lines, whose tracks were also built beside the river. Industry profoundly changed the Susquehanna in the nineteenth century, as newly built factories hijacked the river's power with dams and wheels.

As we paddled out into the river, we saw—in the distance, on the far bank—the smokestacks and hulking rusty buildings of Steelton, a steel plant on the site of one of America's first steel mills. The factory's outfall pipes, discharging into the river, don't scare away hard-core anglers like Veruete and Little. "Just down the river is Three Mile Island, and that's where I go to fish when it ices up," Veruete said of the nuclear power plant. "The fish are attracted by the warm water discharged from the plant. The warmth keeps the river from freezing."

A more significant problem for fish than radioactive waste are the massive dams that were built up and down the river to form artificial lakes for industries, especially for the generation of hydroelectric power. These dams block the passage of migratory fish species, including American shad. Bloated reservoirs along the river are followed by exposed stretches of riverbed cobbled with rubble and mangy weeds. The dams trap tons of sediment and pollution, with unintended consequences. Over the decades, so much dirt and fertilizer from farms—about 185 million tons—has piled up behind the southernmost of the hydroelectric dams, the Conowingo, that the muck overflows its confines and smothers the bay whenever there is a storm.[3]

The Susquehanna River is in failing health, and it's mostly because of centuries of abuse by Pennsylvania and the towns, mills, and farms that surround it. An environmental organization, American Rivers, named the Susquehanna one of "America's Most Endangered Rivers" in 2005, 2011, and 2016.[4]

In some ways, however, the Susquehanna River is better off today than it was a century ago. In 1904, during the height of coal-mining and lumber booms in Pennsylvania, Peter Roberts, a preacher and sociologist, described a section of the river in grim terms. "The fish are exterminated, and the forests, where solemn grandeur and majesty once impressed the souls of men, are no more," Roberts wrote.[5] In the early twentieth century, coal silt—25 feet deep in places—blanketed the bottom of the Susquehanna. Acidic water draining from mines caused massive fish kills and stained streams orange. Raw sewage was a huge problem, with the state's capital, Harrisburg, piping human waste straight into the Susquehanna, untreated, until 1962, when the city built its first sewage treatment plant. Boaters in the 1960s complained that their anchor lines would come up coated with human feces. The river improved during the 1980s and 1990s with the construction of somewhat improved sewage treatment plants. But Harrisburg and dozens of other towns along the river still have grossly inadequate sewer pipe systems that combine human waste with rainwater and become overwhelmed during storms. Whenever even a half inch of rain falls, dozens of outfall pipes in Harrisburg alone still release raw sewage mixed with rainwater into the river.

This sewage causes a bacteria problem that often makes it unhealthy to swim in the Susquehanna. But overall, the river's biggest issue is the runoff of sediment and fertilizer from farms, which is causing tremendous damage downstream in the Chesapeake Bay. Starting in the 1950s, many farmers in Lancaster County—the most intensively farmed area along the Susquehanna—followed a trend nationally and switched from small-scale crop farms to intensive production of large numbers of hogs and cows in industrial-scale metal buildings. These factory farms generate millions of tons of manure—far more than the surrounding farmland can absorb as fertilizer, leading to runoff into the Susquehanna. Worse yet, Pennsylvania exempted its biggest polluters—the farm and coal industries—from many requirements of the state's Clean Streams Law, which was passed in 1937. The Pennsylvania General Assembly in 1980, for example, made it illegal for state or local officials to require farmers to fence their cows out of streams, although cattle defecating in streams and trampling riverbanks is a major source of bacteria and sediment in tributaries to the Susquehanna.[6]

The most popular sport fish in the Susquehanna—smallmouth bass—has been in sharp decline over the past decade, with many of the fish suffering from lesions or deformed sexual organs. Eggs grow in the testes of many male fish. Researchers suspect that a combination of factors is causing the deformities—including endocrine-disrupting chemicals from farm manure and agricultural pesticides. Despite the disease in the fish and frequent algal blooms in summer, the Pennsylvania Department of Environmental Protection year after year refuses to designate the river as officially "impaired" with pollution under the terms of the federal Clean Water Act. This designation might require the farms, sewage plants, and industries that surround the river to do more to clean it up.[7]

"The pollution problem on the river is like a perfect storm," Veruete said, as he cast a fishing line into the dark water. "Every guy who's been on the river for an extended period of time will say he feels like he's catching a lot fewer fish than he used to."

But the river still has some life left in it. From across the water, Jeff Little suddenly screamed. "Yeah! *Whoo hoo!* Fish on! *Fish on!*"

"Already?" Veruete replied, paddling frantically over toward his friend. "Is it a good one? Let's take a look at it."

After a tense, thrashing fight, Little hauled out of the water a tiger-striped fish, greenish bronze, that was about 20 inches long.

"These smallmouth bass are just beautiful," Veruete said, admiring the fish before gently releasing it back into the river. "They are like little tanks that have the engines of sports cars in them."

An hour later, he snagged another smallmouth bass, reeling it in during a fierce battle. He measured the fish—it was about 17 inches long—and photographed it. "Goodbye, buddy," Veruete said, before dropping it back into the steel-gray river. "Oh man, I love that feeling. It makes you feel warm. Even though the water is like 38 degrees, I feel warm now."

"Yeah, I feel it radiating all over," Little said. "Let's get some more."

The Susquehanna is worth saving because of people like Veruete and Little, whose lives and livelihoods are forever caught in the river's flow.

GUNPOWDER RIVER
Development Patterns, as Reflected in the Water

I T WAS A HAZY August day, and on choppy waters of the Gunpowder River northeast of Baltimore, scientists from the Maryland Department of Natural Resources were working in a boat, hauling in a long net. They dumped their catch into a metal tub. Biologist Alexis Park lifted the fish, examined them carefully, and then called to a colleague holding a clipboard. "Five young-of-the-year white perch," Park announced, as she grabbed the wriggling fish and tossed them back into the river. "Five bay anchovies. Brown bullhead. Another young-of-the-year white perch."

Plop . . . plop . . . plop . . . the fish splashed into the water, quickly disappearing into the green-gray murk.

Jim Uphoff, a biologist with the state wildlife agency who leads Maryland's annual fish habitat survey, reached into the bucket and snatched a specimen with a spiny dorsal fin.[1] "It's a beautiful fish, the yellow perch," said Uphoff, holding the fish carefully to avoid being pricked. "It's kinda greenish—light green with dark green bars—and it's got these bright orange fins on the bottom. They get up to 13 or 14 inches long, and," he paused for a moment to lend scientific gravity to his words, "they are *absolutely delicious.*"

Yellow perch are a sentinel of spring in the Chesapeake Bay region, the first species to migrate up streams to spawn. The females lay eggs in long, golden strands that look like silk stockings dangling into waterways from tree roots and branches. These golden ribbons of life, however, are being turned into ghostlike strands of dead eggs in parts of the state with rising densities of human population. Uphoff and his fellow scientists discovered that many of the fish eggs have been failing where suburban growth is more concentrated.

"The eggs just won't hatch," Uphoff said, before gently lowering his hand into the river and releasing the yellow perch. "The historic record in the Severn River, for example, indicated we would see a hatching rate between 80 and 90 percent. But now far less than 10 percent are hatching."

This dramatic drop happened over the past half century, coinciding with the march of sprawl across the Chesapeake Bay region. The human population across the bay watershed has more than doubled since 1950, rising from about 8 million that year to about 18.1 million in 2016—and the numbers are expected to keep rising to 20 million by 2030, according to the EPA Chesapeake Bay Program.[2] Maryland's population growth has been even faster, tripling since World War II, to an estimated 6 million people in 2015.

The patterns of growth have not been evenly spread, however, in Maryland or elsewhere across the United States. Since the development of the interstate road system in the 1950s, development has increasingly leapfrogged outward from cities and towns, with spread-out housing tracts increasingly built in half-rural exurbs, while older communities have lost population. In Maryland, for example, Baltimore bled almost one-third of its population between 1970 and 2010, falling from 905,787 to 620,961 before leveling off to an estimated 625,000 in 2015, according to figures from the Maryland Department of Planning.[3] With this flight from urban areas, the populations of suburban Howard and Calvert Counties more than quadrupled from 1970 to 2010; the numbers in exurban Frederick and Charles Counties nearly tripled; Carroll and Harford Counties more than doubled; and on the Eastern Shore, Worcester and Queen Anne's Counties more than doubled.

This pattern has spread blacktop for malls, roads, and parking lots to replace fields and forests at a rate far more rapid than the rate of population growth. Between 1973 and 2010, the population in Maryland grew by 39 percent, while the amount of land covered by parking lots, roofs, and other development skyrocketed by 154 percent.[4] Meanwhile, every year in the Chesapeake Bay watershed between 25,000 and 36,000 acres of trees are cleared—an expanse of land the size of Baltimore being stripped of its pollution filters every two years.[5] The implications of this development pattern are profound. The impacts include not only worsening water pollution but also longer commutes (and thus more air pollution from vehicles) and reduced public health (because walking to the store is impossible in exurban enclaves).

Uphoff and his fellow biologists focused only on one small corner of this picture, examining fish populations in parts of Maryland with differing intensities of suburban sprawl. The researchers found that a tipping point for fish survival came when about 10 percent of the land in the drainage area of a river or stream is covered in roads, roofs, and parking lots.[6]

In the Severn River area of Anne Arundel County, for example, more than twice this "tipping point"—at least 20 percent of the land—is covered in blacktop or buildings. The consequence has not been good for fish. None of the yellow perch eggs here examined by researchers with the US Geological Survey and the US Fish and Wildlife Service had developed properly. By contrast, the fish eggs were healthy in places like the Choptank River on the Eastern Shore, where only 2 percent of the land is paved or roofed.

Vicki Blazer, a fish biologist with the US Geological Survey, discovered that the yolks of eggs broke apart in waterways that were surrounded by more intense real estate development. Other yolks in suburban streams had strange growths in them. Often, the long, golden strands of membrane that are supposed to protect the eggs were too thin to keep the embryos alive.

What *exactly* is deforming and killing the fish eggs? Uphoff said he does not know for certain, but that research by Blazer's team suggests that likely suspects are endocrine-disrupting chemicals that flow out of sewage treatment plants and septic tanks and off of lawns in suburban subdivisions. These pollutants may be throwing off the hormonal balance of fish and therefore damaging their reproductive systems and eggs.

"The chemicals are in things like personal hygiene products, pesticides, medications, and pharmaceuticals," Uphoff said. "It's this soup that comes out of people's septic systems or sewers, and it is not treated by standard wastewater treatment plants. Basically, the fish are swimming around in this cocktail of drugs and chemicals most of their lives, and they are accumulating enough of them that it's affecting their eggs."

What is the solution? Either more restrained use of these chemicals, or an extremely expensive new generation of sewage treatment plants with advanced systems that employ "reverse osmosis" technology to purify water by filtering it through layers of membranes. Less expensive, and more important in the long run, would simply be better planning for development that leaves more of the Chesapeake's landscape green and more of its rivers golden with perch eggs in the spring.

CORSICA RIVER

A Murky River Experiment's Clear Lessons

O N   A   S U N N Y   M O R N I N G under a blue sky with wispy clouds, I headed
out onto the Corsica River in a motorboat driven by Frank DiGialleonardo. He's president of Corsica River Conservancy, which works to
protect the 6-mile-long waterway on Maryland's Eastern Shore. Although
the river was a sickly olive brown, the air above featured a riotous festival
of bird life. Osprey plunged for fish as blue heron strode through the
shallows and a pair of terns swooped after each other. The banks of the
river were lined with big homes with wide lawns and piers sporting yachts.
Beyond the houses and their stands of shady oak, the landscape faded into
miles of cornfields and poultry farms.

I was on the river that day because 10 years earlier, in 2005, Maryland
governor Robert Ehrlich, frustrated by the slow pace of the Chesapeake
Bay cleanup and facing difficult reelection odds, launched on the Corsica
what he touted as the state's grand new experiment in bay restoration.

Ehrlich promised to concentrate $20 million into an array of coordi-
nated water pollution control projects to try to create a dramatic improvement on one small river, the Corsica.[1] The idea was that if the bay region
was struggling with cleaning up the whole Chesapeake, perhaps Maryland could figure out what would work to quickly improve the health of
one waterway within its own boundaries. Then the results could be rep-
licated elsewhere.

"What I loved about this project, maybe as a lawyer," Ehrlich told me
shortly before my boat trip, "was that we could isolate one river and bring
in all the best practices, and every level of government and nonprofit
organizations, to focus on what works and what doesn't work. What really appealed to me, most of all, was that we could measure it."[2]

Now, some folks dismissed the Corsica project as a political ploy—a head fake to divert attention away from Ehrlich's shortcomings in other areas of environmental policy. A major sewage treatment plant upgrade on the Corsica, for example, had already been planned long before Ehrlich's press conference to announce the Corsica project.

But I thought, we should accept Ehrlich's challenge. Let's *measure it.* And so 10 years after his Corsica River announcement, I decided it was time to make a numeric assessment of the project's success. The bottom line: the Corsica project achieved about two-thirds (10 of 15) of its goals.[3] But this should be taken with a grain of salt because the project's organizers changed many of the goals during the course of the project, so the *actual* success rate remains as murky as the river itself. Most importantly, the restoration effort fell short of its ultimate target of improving water quality in the main section of the Corsica River. The reasons for this muddy and unsatisfying result provide lessons that can be applied to the larger Chesapeake Bay cleanup.

"This was a state project that said, 'We are going to spend $20 million over five years and expect to see results,'" DiGialleonardo said, as we rumbled down the river, past a beautiful 1920s-era Chesapeake Bay workboat called a Hooper's Island Draketail. "But $20 million did not get spent in that five year period. It was closer to $5 or $6 million that was expended. Clearly, it was going to take a lot more than five years to complete the project."

Despite the severe cuts to funding—which, in the end, totaled about $10 million over 10 years—some of the projects were effective. The state modernized the half-century-old wastewater treatment plant in Centreville, which had a history of unreported sewage spills.[4] The Maryland Department of the Environment contributed $4.5 million toward the modernization of the Centreville Wastewater Treatment Plant. This upgrade virtually eliminated the high levels of fecal bacteria from human waste in the river and reduced 50,000 pounds per year of nitrogen pollution.

The state also paid farmers $45 per acre to plant wheat, rye, and other crops without fertilizer in the fall and winter around the river's watershed. Farmers planted these pollution-absorbing cover crops on 5,000 acres, which exceeded the program's goal of 3,000 acres.

Local residents were enthusiastic about installing 500 rain gardens to catch stormwater runoff from their roofs, which was far more than the goal of 200 gardens. And dozens of waterfront property owners started growing oysters in cages next to their piers and then planting the oysters in a sanctuary in the river. State biologists planted about 50 million oysters on 10 acres in the river, which meant that half of the project's goal of planting 100 million baby oysters was realized. The project's organizers aspired to replant 50 acres of wetlands along the river but achieved much more than that, actually restoring 80 acres.

The result of all these projects was positive. Two of the three streams that feed into the Corsica River experienced declines in nitrogen pollution of about 25 percent over the decade, according to monitoring by the Maryland Department of the Environment.[5]

But in the main stem of the river, water quality did not improve over the decade, according to state data.[6] And because the water still remains so murky from algae and sediment, no underwater grasses can survive. The project had a goal of establishing 10 acres of healthy underwater grasses over a decade, but none grew, because the water quality remains so poor.

"With underwater vegetation, our progress has been nil," DiGialleonardo said, gazing down into the pea soup–like water. "The problem is that underwater grasses will die without sunlight. We don't have sufficient sunlight in the water to sustain any underwater vegetation."

This is a problem across the Chesapeake Bay, as water clarity has worsened over the past three decades, falling from a rating of 41 out of 100 in 1986 to an abysmal 6 in 2011, then rising to a 24 in 2016, according to water quality monitoring analyzed by the University of Maryland Center for Environmental Science.[7] The amount of algae in the bay, as indicated by measurements of chlorophyll a (a green pigment in phytoplankton), also increased from 1986 through 2011, as nitrogen and phosphorus pollution fed blooms, which blocked the light.

So why didn't the decade-long project on the Corsica work to make the water clearer? Part of the problem could be that farmers did not cooperate by volunteering to plant trees along the streams on their land, even though forested buffers like this serve as natural filters to catch pollution. Why not? Because farmers know they can make more money planting corn on the land beside their streams. The rising profitability of corn is a major environmental problem. A federal law mandates the production of corn-based ethanol, which is driving up the price of corn. This makes farmers want to plant more acres—even though corn is perhaps the worst crop in terms of creating water pollution, because it requires large amounts of nitrogen fertilizer, which is flushed by rain into streams. Some farmers also don't want to plant trees because they find them inconvenient for the maneuvering of large farm equipment.

The Corsica River restoration project had a goal of using financial incentives to encourage farmers and others to plant 200 acres of trees along streams. But in the end, only 10 acres of these forested buffers were ever planted—and almost all of these trees were planted on government land, not farmland.

"We had a very difficult time with that," admitted Kevin Smith, assistant director of the Chesapeake and Coastal Service at the Maryland Department of Natural Resources, in reference to the lack of participation by farmers in forested buffer programs.[8] "We place a really high value on forested buffers, in part because they provide shade to our waterways, and high water temperatures are hard on aquatic organisms. But farmers prefer grass buffers. Grass is something they can drive their tractors over."

Another area of shortfall was in the control of stormwater runoff pollution in the paved areas of the town of Centreville. The Corsica River project established a goal of installing ditches with pollution-absorbing plants, rock beds, and perforated pipes to filter the water flowing off of 300 acres of parking lots. But in the end, only about one-third of these filtration systems were built, in part because of the cost, which ran up to $15,000 per acre.

Homeowners also were reluctant to volunteer to install upgrades to their septic systems, which are tanks buried under their yards that hold the waste from their toilets. Because these tanks leak nitrogen pollution into the river, the Corsica River restoration program had a goal of upgrading the septic systems of 30 homes with advanced nitrogen removal equipment. But at the end of five years, only 16 homeowners had volunteered. By 2014, it was 23.

Why were people so slow to fix their leaky septic systems? Because of the cost and perceived hassle, which included digging up parts of their lawns. Maryland has a program that pays about $13,000 toward the installation of advanced nitrogen pollution removal equipment in home septic systems. But the reimbursements sometimes are thousands of dollars short of the real cost of the equipment, which can hit $20,000 per home. "People often asked questions along the lines of, 'If my septic system is working, why do I need to change it?'" said Smith. "What it always comes down to, in most cases, is cost."

Finally, there was the most important factor: people. The population of Centreville, at the heart of the Corsica River cleanup project, almost doubled to about 4,500 during the decade when the water pollution control effort was under way. More people meant more toilets being flushed, more car exhaust, and more blacktop.

So what are the lessons for the broader Chesapeake Bay cleanup which we can learn from Governor Ehrlich's little experiment? One clear lesson is that modernizing sewage treatment plants works and is a valuable investment of taxpayer money. In theory, paying farmers to reduce their runoff pollution *might* help, as long as the payments to farmers (in taxpayer dollars) are high enough and the completion and effectiveness of the projects can be verified.

But, in general, purely voluntary efforts—like trying to convince farmers to plant trees, or attempting to persuade homeowners to upgrade their septic system—usually fail. In these cases, government mandates will be required to make real improvements in water quality.

And beyond all this, the single most challenging problem is population growth. If the number of people living in the Chesapeake Bay watershed continues to expand at a rapid rate in patterns that devour trees and fields—as they did in the Corsica River watershed—all our efforts to curb pollution will be washed away.

PATUXENT RIVER

The Patuxent River Will Break Your Heart

F RED TUTMAN was cruising down the Patuxent River in southern Maryland when he entered Jug Bay. It's a lakelike swelling of the river fringed by a field of yellow pond lilies with heart-shaped leaves. As I peered over the edge of his boat, I saw bright flowers on stalks gazing up at me like eyes just beneath the water's surface. Nearby on shore, the trees were painted light green with the unfolding leaves of spring.

"A lot of people think of the Patuxent River as the Chesapeake Bay in miniature," said Tutman, the Patuxent Riverkeeper, as he piloted his boat through the foot-deep water.[1] "The Patuxent is about 110 miles from end to end—the longest river that is located entirely in Maryland. And its watershed has a little bit of all of the ecology, in terms of land use and terrain, that you will find virtually anywhere else in the Chesapeake. It has suburban areas, a few remaining farms, and a lot of marshes and wetlands that are like the kidneys of the watershed. When you remove those wetlands, you remove the ability of the waterway to clean itself."

A wind blew the tall reeds along the shore, making them rattle and sway. Red-winged blackbirds darted among the tufted heads.

"If we lose the Patuxent, we here in Maryland have no one to blame but ourselves," Tutman continued. "If we can't fix this river, it's hopeless that we are ever going to fix the Chesapeake Bay."

Tutman, 57, has been working as the Patuxent Riverkeeper for 13 years. His job, as director of a nonprofit organization of the same name, is to be a champion of the river through educating the public, advocacy, and legal action. "I try to rally people, build enthusiasm for the river, and empower people to fight for water quality," Tutman said.

Before founding the Riverkeeper organization in 2004, Tutman ran a media company for more than 25 years, writing and narrating for TV and working as a videographer and producer for CNN, ABC, NBC, CBC, and BBC. "The problem with that job was that I didn't feel like I was making much of an impact, working behind the camera," Tutman recalled. "I just felt like I was reporting on misery, but never doing anything about it. So I decided to move to a different level and become an advocate, and the environment was what I cared most about."

Tutman grew up beside the river with a first name that pointed toward boldness. His parents chose "Frederick" because they were inspired by the great Maryland abolitionist Frederick Douglass. Tutman grew to be nearly as fiery, outspoken, and driven by moral purpose as his namesake. He is one of a very few African Americans directing an environmental organization in the Chesapeake Bay region or anywhere in the United States. He was inspired to devote his heart and soul to the river by his family's deep roots along the Patuxent, where his family has owned a farm for seven generations.

"We grow hardship, like most American farmers," Tutman chuckled, as he raised an oar to shove his boat out of the motor-clogging mud in a shallow spot in the river. "At the moment, our farm is in an organic transition, moving away from GMO [genetically modified] crops and government-subsidized agriculture, such as growing corn and soybeans. And so right now we're growing organic rye and garlic."

He started playing in the Patuxent as a child and found his imagination hooked forever. He recalled:

When I was a kid, I just loved the river. I was endlessly fascinated. We made rafts. We crabbed. We caught fish. Our favorite way to catch fish was actually to wait until after a big flood, when the fish were trapped in the shallows, up in a mud plain, and we'd scoop them up with buckets. We used to scoop up tons of shad that way. Literally, trash cans full of shad. We hosed out big aluminum cans to fill them with fish, which we scaled and fried sometimes right next to the river. Sometimes, we froze them in freezers. But mostly, I was fascinated by this mysterious force that came from somewhere upriver and went somewhere downriver. The river was like this eternal sense of intrigue. Where did it come from, and where is it going?

The river's destination appears to be a dark place.

In the distance, at the far end of Jug Bay, the smokestacks of Maryland's largest coal-fired power plant rise up. As we motored toward the

Chalk Point Generating Station, it was almost as if everything around us—the river, the wooded banks, the pathway of water—was drawn on a conveyor belt toward this hulking monstrosity. The power plant looked like a castle with black towers looming over the field of yellow lilies.

"Chalk Point is a 1960s vintage coal- and oil-burning power plant," Tutman explained, gazing up at the massive industrial complex. "Back in 2000, a pipeline leading into Chalk Point spilled about 126,000 gallons of oil into the Patuxent in the worst ecological disaster in Maryland's history. And the plant still pollutes at about 40 times its permitted limits out of a discharge pipe into the river. So we have a legal action going on right now against Chalk Point to bring the plant into compliance and to impose fines for its over-the-top discharges."[2]

Tutman said that this legal battle is one of about 10 he is waging against polluters along the river. And while that number may sound excessive, he argued that—historically—lawsuits have been the only thing that has worked to clean up the river.

The river has a fascinating history. In the 1700s, the lands surrounding the waterway were stripped of their trees for the planting of tobacco plantations.[3] For decades, the Patuxent was a highway of commerce, bearing ships packed with profitable pipe weed from the town of Port Tobacco to Europe and beyond. But by 1759, erosion from the tobacco farms dumped so much silt into the Patuxent that large sections of the river were no longer navigable. The gateway to once-prosperous Port Tobacco became a swamp choked with bushes and reeds, and the town itself became a backwater.[4] In the War of 1812, American commodore Joshua Barney, commanding 500 men and a small flotilla of gunboats and ragged barges, fought off a far better armed British Navy squadron of more than 40 ships on the Patuxent River. Commodore Barney won the "Battle of the Barges" by temporarily slowing the British advance. But then, nearly two months later, when the British returned, he was so overwhelmed that he burned his fleet in a vain attempt to block the invading ships. That didn't work, and the British landed at the town of Benedict and marched toward Washington, DC. Barney and his men fought ferociously to try to stop the redcoats, but Barney was wounded and captured. From their river landing on the Patuxent, the British conquered Washington, DC, and set fire to the White House and Capitol.

For much of the century and a half that followed, the Patuxent was a sleepy corner of nowhere, surrounded mostly by tobacco barns, fields, and forests. The Patuxent's largest "city" was at its deep southern end—the quaint sailing metropolis of Solomons (population: 2,300). There, in 1925, the University of Maryland opened the Chesapeake Biological Lab, the oldest continually operating marine laboratory in the East. The land around the Patuxent River experienced a radical transformation in the decades after World War II, when suburban growth from Washington, DC, increased the flow of raw sewage and sediment from construction sites into the river. Each additional resident moving into the Patuxent's drainage area meant another 100 gallons of sewage per day (or 36,500 gallons per year) that had to be treated, and armies of these newcomers were swarming in.[5] By 1977, the *Baltimore Sun* declared, "Steeped in sewage and laced with chlorine, the Patuxent, most of it still beautiful to the eye, is a conduit of the toxic waste of suburban growth."

In 1979, three southern Maryland counties filed a federal lawsuit against the state over this pollution. They claimed that sewage from upstream suburban counties was violating the federal Clean Water Act, then only seven years old and relatively untested.[6] The legal battle was led by Bernie Fowler, an avid crabber, Navy veteran, and Calvert County commissioner from Broome's Island. He owned a boat rental company called Bernie's Boats, and he was disturbed that he could no longer see his shoes when he waded out into the river, because every year, the Patuxent got muddier and more choked with algae. Fowler represented the voices of watermen along the Patuxent River, who were increasingly angry that they could no longer harvest oysters and crabs because the water was so contaminated. Fowler's lawsuit was backed by scientists at the Chesapeake Biological Lab, who couldn't bear to watch the death of the river flowing past their front door. The lawsuit resulted in a 1981 settlement, which led to the creation of a consensus-based coalition of seven counties and the state. The coalition set a goal of restoring water quality in the river to what it had been in 1950. And to implement this goal, state lawmakers in 1984 passed the Patuxent Watershed Act and created the Patuxent River Commission, which were supposed to guide and provide oversight to cleanup efforts. These acts succeeded in forcing the upstream counties to modernize many of their 30-plus sewage treatment plants, and these investments improved water quality somewhat in the Patuxent (although the river never returned to what it was in 1950). The river enjoyed an upward swing in the 1980s and 1990s, but then it started slipping backward again in the 2000s, as growing amounts of runoff pollution from subur-

ban sprawl and population growth overwhelmed the sewage plant upgrades. The number of homes built with septic tanks in the Patuxent River
watershed multiplied—increasing their pollution output from 212,000
pounds of nitrogen and phosphorus per year in 1985 to 480,000 pounds
in 2000, according to figures from the Maryland Department of Planning.[7] Water quality in the river stagnated between 2000 and 2010, but it
has since started showing signs of improvement, with its health rating
rising from 20 out of 100 in 2011 to 38 in 2016, according to the University of Maryland's annual report card.[8] (Despite the improvement, the
only rivers in Maryland with lower health ratings were the Patapsco and
Back Rivers in the Baltimore area, which had abysmal scores of 27 out of
100 in 2016.[9])

"The Patuxent River is in trouble, like many of the Chesapeake Bay
tributaries," Tutman said. "It is a death by a thousand cuts. A big problem
is growth, including runoff from construction sites and parking lots. I
think it should build citizen outrage, and that is one of our goals: to build
constructive outrage to demand change."

Tutman has thrown himself body and mind into a personal war to save
the river. He filed a lawsuit against Walmart over the company's proposal
to build a gigantic Super Walmart in a farm field surrounded by family
farms in Bowie, Maryland, an area that drains into a tributary to the
Patuxent River. Tutman challenges proposals for many big-box stores,
because their massive parking lots dump polluted runoff into waterways.
Walmart's race-to-the-bottom capitalism—and import of cheap Chinese-
made goods—also destroys competing local stores and depresses wages.

"These are David and Goliath fights," Tutman said of his one-man
legal battles against Walmart. "People find them daunting, because they
figure, 'What is a little pipsqueak nonprofit organization going to be able
to do where the government has failed?' But the truth is, I believe in the
power of one. I believe in the power of dissent. I believe in the First
Amendment to the Constitution. I believe you, as an engaged citizen, have
a lot of avenues to take in hand the problems of your local community and
your local waterway. We can solve these problems."

Over the years, Tutman's lawsuits against polluters have won $650
million in penalties that have been invested in pollution control projects.
But he was also hit with a defamation lawsuit by a waterfront restaurant
owner who wanted to intimidate and silence him because of Tutman's
complaint to state environmental officials about the restaurant's polluting

of the river.[10] That "SLAPP" lawsuit—which stands for "Strategic Law-suit Against Public Participation"—ultimately failed. In another case, when Tutman filed a legal challenge to a wetlands destruction permit to build a big-box store complex in Largo, Tutman said a supporter of the project physically attacked him by screaming at him, jumping on him, and punching him.[11]

Litigating to defend the river nearly cost Tutman his life. "The very first lawsuit I worked on came within weeks of me taking the job as Riverkeeper," he recalled, thinking back to events in 2005. "And it ended up in a situation where I stayed up many days and nights drafting a legal complaint, and at the end of it, I just collapsed. I found myself in the cardiac ward. I had pushed too hard and really depleted my stamina. They called it a 'cardiac event.' Basically, my heart stopped beating."

Tutman woke up in the hospital and eventually recovered. When his head cleared, he learned that his near-fatal lawsuit against a sewage treatment plant run by the Washington Suburban Sanitary Commission had ended with a major victory for the river. His long hours and stress paid off in a $350 million settlement from the commission which funded repairs and upgrades at the sewage plant on the Patuxent.

But his brush with death also taught him a lesson: he can't be a guardian of only *natural resources*; he must also be a good steward of his own *personal resources*—his time, his energy, his health, and his heart.

"To make the Patuxent River better will take a little bit of help from everyone," Tutman said, looking out over the waters of Jug Bay, where a few fishermen were now casting their lines among the heart-shaped leaves. "Not just money—but participation and the willingness to step up and express dissent. And that's really tricky for some people. I think people are in favor of environmentalism up until the point where it interferes with their everyday life. It is a burden to be vigilant."

POTOMAC RIVER
Black Roses in the Nation's River

I T WAS A STRANGE SIGHT over the side of my canoe as I glided down the Potomac River. I was just west of Sharpsburg, Maryland, and the water was clear. Sun penetrated all the way to the rocks on the bottom, which shimmered gold. Swarms of tiny fish darted in and out of ribbons of grass that had flowers like stars. A smallmouth bass flew through the underwater forest, followed by a fat, whiskered catfish. The scene was so beautiful that it was almost hard to believe: sunshine, silver-frosted clouds heaped in a bright blue sky, and a river surging with life.

I tied my canoe to the knuckle of a tree root and waded to my chest in the galaxy of the flowers.[1] For a while, I floated on my back, letting the current carry me toward the Chesapeake Bay. The forested banks and farmland of West Virginia drifted by on my right, and on my left floated the rocky riverbank below the Chesapeake & Ohio Canal National Historic Park trail in Maryland. I swam back to my canoe and climbed in.

When I paddled on, I saw something disturbing growing from the bottom not far from where I had been swimming. Blobs of sickly neon green and black algae clung to the tops of grass stalks, making the plants look like black roses. Elsewhere, clumps tumbled down the river like fuzzy alien balls. The common name for this stuff is blue-green algae. But it is actually not algae or a plant at all. It's a phylum of bacteria called cyanobacteria, and it is one of the most primitive forms of life on our planet. This pond scum made human life possible by creating Earth's first oxygen. But today, too much cyanobacteria is multiplying in waterways around the world. It is fed by fertilizer that is washed by rain off of farms and lawns, as well as by nitrogen and phosphorus pollution from sewage treatment plants, air pollution, industries, and other sources.

Although pond scum invented the air we breathe, when it dies, it sucks oxygen out of the water, spawning low-oxygen "dead zones" in the Chesapeake Bay and other waterways. The bacteria also produces a toxin that can kill fish and cause skin rashes and vomiting. A cyanobacteria bloom on Lake Erie in 2014 released so many toxins into Toledo's drinking water supply that the city had to temporarily shut down its water treatment plants, forcing people to rely on bottled water. Massive, fish-killing cyanobacteria blooms have also been erupting in Florida, California, Wisconsin, Utah, and many other states in recent years.[2]

I paddled hard to move away from the black roses quickly, but the water never completely cleared up. Occasional globs of the cyanobacteria drifted down the Potomac River for miles. Between the clumps, the jagged edge of a sheet of rock broke the surface, like the knobby spine of a sea serpent swimming up the river. My canoe got hung up between two vertebrae in the spine, and I almost tipped over. But I rocked the boat back and forth with my weight until, eventually, with a hard shove from my paddle, I floated free again. Rapids rushed past, and I found myself drifting beneath a row of white sycamores on the western side of the river.

As I eased around a bend at Shepherdstown, West Virginia, I saw a remarkable sight: the crumbling stone buttress of an old bridge rising from the water, with a twisted tree erupting from the shattered face of the tower. Nearby were two other stone foundations topped with bushes like unruly wigs.

The sun began to set. I dragged my canoe onto a steep muddy bank on the Maryland side of the river near Antietam Creek and set up a tent on a bluff overlooking the Potomac. I sat on the slope and watched a mist rise from the river and stars emerge and brighten across the dark sky. I thought back to the garden I had seen earlier that day: the beautiful yellow flowers I had waded through, and the repugnant black roses.

It occurred to me that the Potomac River, which flows through Washington, DC, is a perfect example of why we need Washington to help with our environmental problems. That may seem like an obvious point. But many politicians of both parties ridicule Washington and are increasingly hostile not only to federal regulation but also to the basic idea of government action to make our world a better place. Some voters throw up their hands in despair over the toxic algal bloom that is national politics these days, particularly the sickening dysfunction of Congress. I happen to love the District of Columbia (where I went to college and often work), but it's come to the point where the word "Washington" has become a political slur thrown around by both parties—far more damaging than "socialist," "tax dodger," or even "racist." But the fact of the matter is that Washington, DC, is the manifestation of Americans' collective effort to manage and improve our world through our democratic system. And from an environmental perspective, federal intervention is absolutely necessary to make the water cleaner when there are so many different sources of pollution spread out over so many states in a multijurisdictional waterway like the Chesapeake Bay. If Washington did not set limits for everyone, self-interest would (and does) poison our waters.

The Chesapeake's problems are so broad ranging and multifaceted that no one person or one state government could ever clean them up. Private enterprise would have no incentive to. As I looked across the river to a farm field beneath the moonlight, I thought: What if I were the owner of that farm? If I decided to reduce my spraying of fertilizers to help the Potomac River, but a subdivision was built downstream that started to spread more lawn fertilizer, the river would become worse, despite my best intentions (and my reduction in income by producing less corn through the application of less fertilizer). If Maryland imposed a tax on its residents to modernize its sewage treatment plants, but West Virginia and Pennsylvania did not and allowed more homes to be built, then Maryland's investments would be flushed down the toilets by its neighbors. This is very close to what happened in 2004, when Maryland created a "flush tax" but these neighboring states did not. Because Maryland's geography and identity are defined by the Chesapeake Bay, the state was motivated in 2012 to also pass a law mandating stormwater pollution control fees (dubbed the "rain tax" by opponents) to pay for the construction of pollution control systems for Baltimore and the state's 10 largest counties. But Pennsylvania, West Virginia, Delaware, and New York did not impose these fees, even though their pollution eventually winds up in Maryland's waters. Maryland's taxpayers rose up in revolt; the rain tax mandate was dropped.

Extreme inequality of effort among bay region states is the dirty reality. Each of the six different states in the Chesapeake region has a distinct political culture, and often opposing political parties in power, making cooperation difficult. Add on top of this the challenge of coordinating with independent local governments that run the water pollution control programs in thousands of counties, cities, towns, and boroughs across the

Chesapeake watershed. Pennsylvania alone has 2,562 different municipalities, many of which control their own separate sewer and stormwater systems and decide whether or not to raise fees to pay for upgrades.

Clearly, strong federal leadership is needed so that each of the state and local governments in this sprawling and chaotic galaxy fairly contributes to solve the water pollution problems they all create. Left on their own, the separate states and communities will just send the problem downstream to foul someone else's water. Likewise, in a purely free-market system, each individual person will tend to pursue whatever is cheapest and easiest for himself or herself. Individually, the private homes, swimming pools, and yards of the wealthy may be beautiful. But our shared spaces suffer. Shared beauty is lost. Nature, which belongs to no one, is destroyed by everyone.

Still, I understand that some people fear and resent federal power—a reality that cannot be avoided when contemplating the history of bloody political conflict over the Potomac River. The 405-mile waterway drains 14,700 square miles of the District of Columbia and Virginia (capitals of the Union and Confederacy), as well as Maryland and West Virginia, pouring about 10 billion gallons of water a day into the Chesapeake Bay.[3] The Potomac—home of George Washington, a shad fisherman—was the flashpoint of clashes between Virginia and Maryland over fishing, oyster harvesting, and water rights stretching back to colonial times.[4] During the Civil War, Harper's Ferry, on the Potomac just downstream from where I camped, was the site of John Brown's raid on the federal arsenal —his attempt to spark an insurrection to end slavery. Confederate general Robert E. Lee's armies crossed the Potomac for the epic battles of Antietam and Gettysburg. And at the mouth of the Potomac is Point Lookout, a finger of land jutting out into the Chesapeake Bay, on which the Union built the largest prison camp of the Civil War, Camp Hoffman, where 4,000 Confederates died and were dumped in a mass grave in a conflict over the whole idea of Washington, DC, telling them what to do. One might have hoped that all those deaths would have settled the question of federal versus state power, but they did not—as that political war continues to rage on, bleeding even into our waters and air.

Although the Potomac runs through our nation's history, for a long time "America's River" was a disgrace. During much of the twentieth century, the Potomac was literally the sewer of the District of Columbia, with feces from the White House and Congress and everywhere else in the city pouring directly into the reeking river. In 1957, the US Public Health Service declared the river unsafe for human contact. In the summertime in the 1960s and 1970s, blankets of toxic algae often smothered the Potomac. The DC region's sewage treatment plant, called Blue Plains, was built in 1938, but it was grossly inadequate until a major expansion and upgrade from 1970 to 1983 installed more advanced pollution removal systems. The upgrades—which cost more than $1 billion—were forced in part by Congress's passage of the federal Clean Water Act of 1972. In 1996, Blue Plains was modified again to cut its nitrogen pollution by another 40 percent. The city's sewage system and Blue Plains are now being rebuilt once more—at a cost of another $1 billion plus—in a massive project to further reduce pollution and stop raw sewage overflows from city pipes during rainstorms. The impact of the investments in sewage treatment over the past three decades has been obvious and impressive. Water quality in the Potomac improved dramatically in the 1990s and 2000s, allowing the return of lush forests of underwater grasses to a river that had been barren for decades. With the grasses returned waterfowl that feed on the plants, as well as fish that need the grasses as habitat. National bass fishing tournaments are now held on the Potomac River, which was unthinkable a half century ago. The work is far from done, with sewage plants and farms upstream from the district still a problem and sewage leaks from the city still continuing (although being addressed by a massive tunnel project and other improvements currently under construction).

Still, the efforts on the river seem to be progressing, and the waterway is growing healthier. Although Washington, DC, has become the whipping boy of politicians and pundits, the Potomac River embodies the whole point of central government: to bring together the clashing interests of different people and get them all flowing in the same direction.

To those who rage against Washington, I offer them the bouquet of black roses growing at the bottom of the "Nation's River."

JAMES RIVER

The Spirits of the Swamp

T HE SUN WAS SINKING low over the James River when I set out into the waters that flow through the dark heart of American history.

From the roadside where I had parked my car on this September evening, I dragged my kayak through brambles and shoe-sucking mud to launch into the dark, warm current flowing past Jamestown Island in Virginia. As I pulled the thorns from my shirt and paddled over the glassy water, it struck me as downright bizarre that the first English colonists in North America (after the ominous disappearance of the Roanoke settlers) would choose this briar-ridden, brackish swamp as the home base for their dreams. Of course, this original American Dream quickly turned nightmarish for Captain Christopher Newport's 104 upper-crust adventurers. Their lack of good judgment in real estate combined with a lack of potable water in this marshland to produce illness, starvation, cannibalism, murder, and an escalating war of ethnic extermination against the far more numerous Algonquian people, led by Chief Powhatan.[1] Of the first 7,200 gentlemen and ladies from England who settled in Jamestown between 1607 and 1624, only 17 percent (or 1,220) survived beyond 1625.[2] What finally saved the for-profit business that was the Virginia colony was a miracle crop: tobacco, tended by African slaves on land stolen from the Native Americans, which quickly addicted the world to its deadly but lucrative pleasures. From this swamp did the American enterprise grow— the least likely possible soil for the flower of democracy.

The serene beauty of these wetlands, with their waters painted by the setting sun, does nothing to betray their violent history. As I dipped my hand into the warm water, I almost forgot that this place has a past. It seemed to offer only a present: a cheerful song of the crickets in the marsh grasses. Beauty has a way of doing this—obscuring the troubled things that lie beneath it. But then I saw something that spoke to me of Jamestown: a huge and ancient bald cypress tree, rising from the still water. I paddled toward where it stood amid a cluster of its gothic brethren. Their wide, flaring bases were like the black robes of a council of elders, with the knees of their roots bulging out of the water beyond their vestments. These gnarled spirits of the swamp, I thought to myself, must have seen it all: the bloodshed and insanity of the birth of America, and then immense changes to the landscape and people over the centuries. I drifted in close and touched the rough skin of a cypress, and it told me what it knew.

The James River, over which the cypresses keep watch, is awash in history. It was originally the Powhatan River, named after the chief who ruled this land until King James's people moved in. The James River was the place where Captain John Smith, the most lowborn of Newport's crew, met Pocahontas, the most noble of Powhatan's empire. It was here that the princess then married John Rolfe, the pioneering drug dealer who invented the tobacco industry and brought his exotic young wife to London, where she died. The James was the port of entry for both slavery and democracy to America. It was the backdrop for Patrick Henry's words "Give me liberty, or give me death!" which stirred a couple of locals named Thomas Jefferson and George Washington. The James was the birthplace of corn, cotton, and iron mills that marked the beginnings of America's industrial revolution. It was the artery of commerce that fed the Confederacy and its heart, Richmond. The James River was the stage of the epic battle between the world's first ironclad warships, the *Monitor* and the *Merrimack*, and the pathway for the siege and burning of Richmond and the fall of the Confederacy. It was up the James River that Abraham Lincoln's boat, the USS *Malvern*, carried the liberator to personally greet the slaves he had freed.

The James is "the river where America began," wrote journalist and environmentalist Bob Deans, in his book of the same title.[3] "I was well into my teens before I realized that the phrase 'Cradle of Civilization' refers to the Tigris-Euphrates river valley in present day Iraq, and not, as I had long presumed, to the confluence of the James and Appomattox rivers between Richmond and Jamestown."

After spending some time with the bald cypresses, I paddled out over the smooth black water to a wide bay. An osprey perched atop a dead tree at the edge, piping piercing notes. I passed beneath a cement bridge and out into the main section of the James River itself.

The sweep of the James here is impressive: almost a mile and a half across, and looking like a highway of water. The James runs 340 miles from the Appalachian Mountains to the southern base of the Chesapeake Bay and drains a watershed of 10,000 square miles that is home to about three million people (including in the Richmond area). Seven billion gallons of water a day pour out of the James into the Chesapeake, making it the third-largest bay tributary, after the Susquehanna and Potomac Rivers. The James is the home to one of the world's largest harbors (at Norfolk), several major military bases, and an entire fleet of mothballed ships, the Navy's James River Reserve Fleet.

As I paddled, a horn announced the presence of the Jamestown-Scotland Ferry, as it churned the waters and carried about 50 cars across the wide river. Its wake tossed me sideways, but I recovered and pressed onward, past a shoreline of tall grass and over an artificial reef of boulders. I made my way into a little alcove, where replicas of ships of the 1606 English expedition, the *Susan Constant*, the *Godspeed*, and the *Discovery*, rose up over me. Although I paddled right under their bows, they didn't impress me with their size. In fact, they struck me as preposterously tiny as the vessels that carried a civilization across the ocean. One of them—the *Discovery*, which later ferried the explorer Henry Hudson to mutiny and death in Hudson Bay—looked scarcely bigger than two suburban minivans.

As I was examining these museum pieces, I heard laughter and spun around to see a motorboat with four people chugging beer as they powered full blast into the river, their stereo cranked up loud and their wakes rolling a line of purple hills across the waterway.

The clash between the James's past and its present can be jarring. Despite its great historical and cultural importance, the James has been dishonored by a horrific amount of pollution. The river's natural history is full of chapters as gruesome as its colonial history. The James was an open sewer for the city of Richmond for two centuries, with raw human waste dumped directly into the waterway, and with the many factories along the river also flushing in their chemicals and garbage. The waters were foul in the nineteenth and early twentieth centuries, with sewage that people could see and smell.

In 1975, the James River acquired the reputation as perhaps the most polluted waterway in the world. A factory owned by Allied Chemical in Hopewell flushed down its storm drains and into the river tons of an insecticide called kepone. The plant also exposed its workers to so much of the now-banned chemical that they suffered from tremors and shakes. In response, Virginia governor Mills Godwin Jr. closed a 100-mile section of the river to fishing, from Richmond to the Chesapeake Bay, and the ban remained in place from 1975 to 1981. The river was reopened without the pesticide ever actually being removed, and a thin layer of sediment is now the only thing covering the toxic brew on the bottom.

"That is a pretty nasty legacy," Jamie Brunkow, the Lower James Riverkeeper, told me during a phone call when I got back on shore.[4] "You hear about the Cuyahoga River [in Ohio] catching on fire, and Lake Erie being a dead ecosystem. But I don't know of any river that was totally shut down like the James. Since then, we like to tell folks that the James is the most improved river in the country."

Water quality in the river slowly improved over a half century with the construction of and upgrades to sewage treatment plants. The first sewage treatment plant on the river was built in the late 1950s, in south Richmond. Then, after the passage of the federal Clean Water Act in 1972, several more wastewater treatment plants were constructed up and down the James. These facilities have been gradually upgraded over the decades, reducing fecal bacteria and the nitrogen and phosphorus pollution that feed algal blooms. From 1985 to about 2000, phosphorus and nitrogen pollution levels declined significantly in the river, according to water quality monitoring by the US Geological Survey.[5] The river's health is still not great—earning a rating of 61 out of a possible 100 on the James River Association's most recent "State of the James" report card.[6] But compared to where it had been, the James has recovered more than any other major river in the Chesapeake Bay watershed, according to the University of Maryland Center for Environmental Science.[7] Among the positive signs for the James are high levels of dissolved oxygen in the river and a resurgence of rare Atlantic sturgeon, whose numbers—although still small—are more than biologists ever thought possible.

"The James River also has more bald eagles now than anywhere else in the Chesapeake Bay watershed, and we are seeing aquatic grasses starting to rebound as well," Brunkow said. "As recently as 2012, there were no underwater grasses in the main stem of the James River. And now we are seeing them grow and spread, and that is a really positive sign—because grasses are a keystone species that holds the sediment in place, provides habitat for critters, and produces oxygen."

Despite the improvements, the longest river in Virginia still has miles to go. Sediment runoff and poor water clarity continue to be problems, especially in the tidal portion of the James, with the waters meeting water quality standards less than 10 percent of the time. Another serious problem in urban areas like Newport News are stormwater outfalls that spew excessive amounts of nitrogen and phosphorus pollution from lawn fertilizers and sewage. In the hot days of late summer, these nutrients feed toxic algal blooms that often force state health officials to temporarily shut down the harvest of oysters.[8]

As night closed in on the river, I paddled back toward my car. Tiny fish began to jump from the water, making it look like it was raining.

As I slid past the bald cypress trees, I half expected the council of elders to give me some kind of prophetic warning—a few words of wisdom to pass on to future generations about all they had seen of human idiocy and power.

But the only voices I heard were of Canada geese as they flew overhead in a V formation—a sign that evenings like this would soon turn cold. And then I saw a bat, flying in a crazy, jerky pattern as the stars brightened in the night sky.

SOUTHERN BAY

A Kayak Expedition to the Mouth of the Chesapeake

WHEN THE GRAY CLOUDS parted and a brilliant blue sky opened up, I saw we had picked perhaps the most beautiful place on Earth to set up camp. A row of tall pine trees towered over a sandy bluff and a stretch of virgin beach lapped by waves. The scene looked as the Chesapeake Bay must have more than 400 years ago, before Europeans arrived.

I had joined a three-day paddling and camping expedition down the lower Eastern Shore of Virginia to the mouth of the bay. The trip had been organized by Chesapeake Bay educator and naturalist Don Baugh, with help from his friend, the veteran bay author Tom Horton, a former colleague of mine at the *Baltimore Sun*.

"We invite people kayaking so they can have their own personal connections to the bay," Baugh said, as his team of paddlers ate breakfast and readied their gear.[1] "What this does for people is that it allows them to plug back into the real Chesapeake Bay. Not the Chesapeake Bay in reports or in data—which is important—but a visceral feeling of connection that really makes people want to bring the bay back."

After packing up, we paddled south along the beaches from Hungar's Creek, near Eastville, Virginia, down toward where the estuary opens to the Atlantic. The shorelines were lush with forests and grassy dunes, but the stretches of beach were broken by occasional new homes with all the trees cut down around them and replaced by carpets of turf grass and walls of rocks to separate them from the bay.

As the sun beat down on us, we passed a waterman who motored from buoy to buoy in a Carolina Skiff called the *Lucky Star*. He lifted crab pots out of the water, shook out his catch, and then refilled the traps with bait before heaving them back in.

The wind was strong at our backs. Drifting past in the warm water were strands of eelgrass, which felt like hair when you touched them. Our boats slid over lines of buoys holding nets that stretched almost a quarter mile into the bay. Watermen had driven irregular poles—the trunks of scrawny pines—into the sand to hold these fishing impoundments in place. Perched atop the poles were double-crested cormorants, which surveyed the waves like professors with graying temples.

History was all around us and beneath us.

Although there was no way to tell from the water, we were paddling over a massive crater left by a meteorite that, 35 million years earlier, had smashed into what later became the lower Chesapeake Bay. The meteor burned everything in its path, vaporizing water, shattering rocks, and blasting fragments thousands of miles as it plunged a mile below the surface.

But on the day I visited the disaster site, the water was so peaceful that it looked like it never could have happened.

Far out on the bay, beneath mountains of silver and white clouds, a massive, box-shaped container ship crept along the horizon's green line. The ship loomed over a small boat with a white sail which looked like an eighteenth-century schooner being eaten by an aircraft carrier.

The ships of the Chesapeake Bay have defined American history.[2] After the English established the colony at Jamestown in 1607, Captain John Smith piloted a small open workboat, called a shallop, up and down the bay's tributaries, searching in vain for a Northwest Passage to India. He didn't find India but drew the first map of the Chesapeake. In 1619, an English warship called the *White Lion* sailed into the Chesapeake carrying the first African slaves into North America, arriving more than a year before the *Mayflower* brought white settlers to Plymouth Rock. On September 5, 1781, a French fleet aligned with the American colonists fought off British warships in the pivotal Battle of the Chesapeake. This prevented the British from resupplying the army of General Charles Cornwallis, which was surrounded by the colonists and their allies at Yorktown, forcing the British to surrender in the American Revolutionary War. In the War of 1812, the infant nation was almost strangled to death in the Chesapeake Bay when a British fleet burned Washington, DC, bombarded Fort McHenry, and marched on Baltimore. But the Baltimoreans (then as now) refused to surrender, and so America exists today. During the Civil War, the Union's assault on Richmond came first from the Chesapeake Bay, and so did the US naval blockade that eventually starved the Confederacy. During World War II, most of the Navy air squadrons that fought to protect the United States and its allies were trained in the southern Chesapeake Bay at Naval Station Norfolk, the largest naval base in the world.

A bit north of this base, our tiny fleet of kayaks cut through the waves as storm clouds mushroomed into the sky above us. We were now south of Cape Charles, and we hauled our boats onto the beach of a sandy, uninhabited island at the mouth of a wandering stream. Old Plantation Creek is the first significant waterway on the bay side of Virginia's Eastern Shore north of the estuary's opening to the Atlantic Ocean. We popped open the sealed wells at the front and back of our kayaks and pulled out our tents and sleeping bags. Then we slogged through marsh grasses and muck—with hundreds of fiddler crabs fleeing from our footsteps in clattering waves—to find bare spots to set up our camping sites.

After arranging my tent, I took a swim in the creek. I savored the cool of the water and the feeling of being swept away by the current. As I drifted, I saw white poles marking the locations of underwater cages in which watermen were growing thousands of clams. A cow-nosed ray grazed among the poles, eating stray shellfish.

Back on shore, I talked with Tom Horton, who sat in a folding chair at the water's edge, admiring the traffic jam of life in the shallows, including slow-moving periwinkles and darting needlefish. "One of the things that makes the bay special is that it's a place where land and water intertwine so extensively and intimately," said Horton, author of *Bay Country*, *Turning the Tide*, *Chesapeake: Bay of Light*, and several other books about the Chesapeake.

"I mean, the bay is only 187 miles long. But if you teased out all of the tidal shorelines, and all the islands and rivers, you would have more than 10,000 miles of waterfront," Horton said. "And that's more than a statistical curiosity. Because these margins of land and water—the mudflats, the seagrass beds, the oyster rocks, the tidal wetlands—are among the most productive habitats on earth. That's a lot of life in a beautiful place."

The bay is a beautiful but troubled place. The Chesapeake has been stuck in a slow downward slide for decades, with its overall health declining from a grade of 48 out of 100 in 1986 to a grade of 40 in 2011, as measured by the University of Maryland Center for Environmental Science's annual report card on the bay.[3] Over the years since then, however, the estuary has enjoyed a slight but encouraging improvement, with its health score rising from a 40 in 2011 to a 54 in 2016 (the second-highest mark on record since 1986).[4] Although the uptick after 2011 may have been influenced by a drop in rainfall over this period (less rain means less runoff pollution into the bay from farms fields and suburban parking lots), the improvement in health seems more significant than just a random change in the weather. Nitrogen pollution in the bay declined impressively between 2011 and 2016, with the rating rising from a 38 (out of 100) to a 55 (with the higher grade indicating lower, and therefore healthier, levels of nitrogen).[5] Water clarity in the bay, after a quarter century of continuous decline, improved markedly, from an abysmal rating of 6 in 2011 to a 24 in 2016. Evidence of algal blooms has diminished over the past five years, while the amount of dissolved oxygen in the water has risen. Meanwhile, bottom-dwelling organisms (like clams and worms) became more numerous from 2011 to 2015, providing more food for crabs and fish.[6] The number of oysters in the bay doubled between 2010 and 2014.[7] Most importantly, underwater grasses in the bay rebounded strongly, increasing from 57,964 acres in 2011 to 97,433 acres in 2016, the largest expanse of aquatic vegetation since monitoring began more than three decades ago, according to the Virginia Institute of Marine Science.[8] To put this good news in perspective, however, the expanse of bay grasses in 2016—while better than in previous years—was still just a fraction of the 600,000 acres that historically filtered and provided oxygen for the bay and less than half of the 185,000 acres set as a goal by the bay region states for 2010.

All of these encouraging changes have happened while the Chesapeake was under a new cleanup plan—the EPA's 2010 pollution "diet" for the bay, called the "Total Maximum Daily Load" (TMDL). Under the plan, the EPA established pollution limits for the bay region states and called for a 25 percent reduction in nitrogen and a 24 percent cut in phosphorus pollution, with the federal agency threatening potential penalties against states that failed to meet their targets. It would be tempting to attribute the recent improvements to these new EPA pollution limits and the TMDL system. But the truth is that many of the improvements actually have roots that stretch back many years before the 2010 plan, including to Maryland governor Robert Ehrlich's 2004 "flush tax," imposing fees on homeowners which paid for upgrades to many sewage treatment plants, and to a federal air pollution control law passed in 1990 which reduced nitrogen oxide carried by rain into the bay.[9]

"We still have a lot of work to do to bring this bay back, to make it healthier," Horton said. "But sometimes I'll go to a place like California, and somebody will say, 'Oh, you're from the Chesapeake? Too bad it died.' And I'll say, 'Whoa! We didn't mean to give you *that* impression.' This bay is still quite productive, by any standard—except that of 50 years or 100 years ago, when it was really productive. The thing you fear is a shifting baseline, where each generation settles for less and less. We've seen that all too often. No, we can't go back to the bay of 1608. But I don't want, a couple generations from now, for people to settle for half of what I'm seeing out here today."

As he spoke, in the shadow of a tree hanging over the water, a snowy egret waded in the shallows, spearing dinner with its long beak.

I asked Horton what would be required to accelerate the bay cleanup. He suggested, first, that more trees should be planted on farmland, to create natural filters to catch fertilizer and soil running off the land before it pollutes waterways. He proposed that farmers should be offered financial incentives to allow a portion of their cropland to revert to forests. "The other thing I would do is begin a serious debate on how to pursue a prosperous economy that's not dependent on adding millions more people every couple decades, and that's not dependent on physically consuming more of our natural resources," Horton said. "We just can't keep going the way we are, with endless economic and population growth, and restore the bay."

Storm clouds mounded up over our campsite, their bottom edges darkening.

Don Baugh shouted from across the island, "We have about 15 minutes before a massive rainstorm rolls through! Get ready! Put everything inside your tents!"

The winds picked up and then slammed the island. People scrambled and took shelter, as the tents were buffeted. But then, as quickly as the winds came, they departed and the clouds broke apart. The sun blazed through, and the threat of rain was gone.

As the setting sun pierced the ragged remains of the clouds, Horton and Baugh built a fire in the sand. In the firelight, Horton pried open oysters and laid them on a log beside slices of lemon and a bowl of cocktail sauce. "For my money, this is the best place in the bay," Horton said, as the tired kayakers gathered around and feasted on oysters and beer.

Baugh slid a pan onto the fire and fried crab cakes in butter. Later, when people were relaxing, he explained the history of the stream beside our campsite. Old Plantation Creek flows past the site of the Eastern Shore's first European farm.

"We know that right off the mouth of this stream, right across from where we are, there was an old Indian village and burial ground," Baugh said. "Unfortunately, when the colonists came, they plopped their homes right on top of where the natives were. On top of that village was built one of the largest plantation homes in America, back in the 1600s."

The next morning, when we awoke on the island, we found that our tents and the surrounding marsh grass were covered with glittering diamonds of water. As the sun rose, a pair of royal terns chased one another, calling to each other with piercing cries.

We packed up and ate a breakfast of oatmeal, dried fruit, and coffee. Then we climbed back into our kayaks and continued our paddle south along the coastline.

Here and there, trees had tumbled from the sandy cliffs onto the beach and were jutting out into the waves. At one location, we cruised past a long row of cement blocks that a builder had dumped into the bay to try to stop the rising sea levels from threatening a new mansion. A large sign on the beach proclaiming "No trespassing!" was about to be swallowed by the bay.

A few miles south, off Kiptopeke State Park, we glided into a strange landscape. Nine World War II–era cargo ships were chained together offshore, rusting and falling apart and sprouting tufts of grass atop their weather-beaten decks and roofs.

Upon closer inspection, these vessels turned out to be made not of steel, but of cement—an odd material with which to build a boat, but one that was used during the war because it was cheap. After the war ended, engineers intentionally sunk the ships off Kiptopeke to create a breakwater to protect a nearby pier and ferry service.

As I paddled around the wrecks from a bygone era, I saw a dolphin leap from the water, its back flashing in the sun. I thought to myself, there is a lot of magic still left in the Chesapeake Bay. But it could disappear as quietly as a fin in the waves.

The People

I N THIS GROUP of chapters, I will introduce you to seven people whose lives have been important to the Chesapeake Bay in different ways.

The first two are Maryland governors—Harry Hughes and Parris Glendening—whose combined 16 years in office were critical to the launch and fate of the bay restoration effort.

Working quietly behind the scenes for both was John Griffin, a career natural resources manager who—oddly enough—received his spiritual dedication to the Chesapeake from an alien land.

Next, I profile two local volunteers, Bonnie Bick and Michael Beer, whose lives I honor as examples of how individual citizens, without money or large organizations, can make a profound difference in protecting nature and inspiring others.

You will meet an Eastern Shore poultry producer, Carole Morison, who liberated herself from Perdue's factory model of contract farming and found a more humane and profitable way to grow.

And finally, you'll spend time with a waterman, James "Ooker" Eskridge, the Republican mayor of Tangier Island, Virginia, who is fighting heroically to save his fishing village from a catastrophe that he doesn't even believe in.

HARRY HUGHES

*The Unexpected Captain
of the Bay Cleanup*

H ARRY HUGHES NEVER THOUGHT he'd be remembered for launching
the effort to clean up the Chesapeake Bay. He thought he'd be immortalized for his pitching arm. Growing up on Maryland's Eastern
Shore, his most cherished childhood memory was receiving a baseball
glove as a birthday present when he was six. From that moment on, his
goal in life was to be a professional baseball player—and he got pretty far.
He was a starting pitcher for the University of Maryland's baseball team,
and later he was hired by the New York Yankees to play for their farm
team in Easton, Maryland. But after getting paid only $150 per month,
his future wife, Pat—who sat in the stands, reading books—told him to
take a swing at law school.

"If I'd stayed in baseball, I'm not sure my life would have been too
good, frankly," Hughes recalled at the age of 88 while sitting in a wing-
backed chair at his home in Denton, Maryland.[1] "I was 24 at the time, and
if I'd fooled around in the minors for a few more years, I'd be 28 or 29 and
without any real long-term prospects for a job. I just didn't think that was
worth it. Remember, that was *class* D baseball, the lowest league there
was. They don't even *have* a class D now. I'm not sure I was all that good,"
he laughed. "Things turned out alright anyway, I guess."

After earning his law degree, in 1952, Hughes was elected to the
Maryland House of Delegates and then to the state senate, representing his native Caroline County.[2] He served as state transportation secretary, and then he was elected governor in 1978 in a shocking upset. In a
crowded Democratic primary, Hughes was a little-known Eastern Shore
outsider—mocked as a hayseed, or a "lost ball in high grass," by one
opponent—up against a powerful Democratic establishment candidate,

the acting governor, Blair Lee III, whose administration was shadowed by corruption. Lee became the state's top executive when Governor Marvin Mandel resigned because of racketeering and mail fraud convictions. Mandel was Maryland's second consecutive governor to be convicted of corruption, after Spiro T. Agnew. Hughes's plan was to make his mark on history by cleaning up Maryland's government. Instead, he launched a far more difficult crusade to clean up the nation's largest estuary. He didn't invent the idea of saving the bay—he'd been inspired by activists, writers, and a few politicians who had been ringing the alarm bell for a decade. Notable among the bell ringers was US senator Charles "Mac" Mathias, a liberal Republican from Western Maryland who undertook a five-day, 450-mile boat trip to examine the bay's problems firsthand in 1973 and then secured federal funding for a $27 million EPA study of bay pollution.[3] But Hughes was the first governor to take up the cause, and he coordinated a multistate partnership with the federal government to transform words into action to attack bay pollution.

"By the time I had become governor, it was obvious to anyone who had grown up around the bay that something was seriously wrong," Hughes wrote in his 2006 autobiography, *My Unexpected Journey*. He continued,

> For the three and a half centuries since Europeans arrived, the Chesapeake had been one of the most productive estuaries in the world. But by the latter part of the 1970's, the Bay could no longer produce crabs, fish, or oysters in the quantities Marylanders and Virginians alike had come to expect. Oyster populations were crashing. For the first time in Bay history, the value of the blue crab catch exceeded that of oysters, a pattern that has only widened. Landings of the most valuable and sought-after finfish in the Bay, the rockfish, were also declining. Scientists had begun to document a substantial decline in underwater grasses.

Hughes wasn't interested in the bay because of his upbringing. Although he grew up on the Eastern Shore and did some crabbing as a kid, his family (he was raised by his mother, a teacher) did not have enough money for a boat. Once he was governor, however, he began to appreciate the centrality of the Chesapeake to the state's identity as the "Land of Pleasant Living" (to steal the slogan of a National Bohemian beer commercial that Hughes said he loved). And Hughes was inspired to action by the 1983 EPA report (launched by Senator Mathias) on the growth of low-oxygen "dead zones" in the Chesapeake Bay fueled by phosphorus and nitrogen pollution.

"When that federal report came out, I got my people together and said, 'Let's do something about this,'" Hughes recalled. "A lot of studies had been done previously and had just been put on the shelf and forgotten about. And I said, 'Let's not forget about this one. Let's actually do something about it.' And so that's how we started the Chesapeake Bay Program. The bay, after all, runs right down the center of Maryland and it's a marvelous natural resource—very unusual among all of the bays of the world. It was a very easy decision, as far as I was concerned, that we ought to do whatever we can to preserve it."

In the 1983 state legislative session, Hughes asked the Maryland General Assembly for funding to get the ball rolling on the bay cleanup.[4] He and his cabinet organized a multistate partnership to restore the nation's largest estuary, which was the first regional environmental effort of its kind in the nation. Hughes called and wrote the governors of Pennsylvania (Richard Thornburg) and Virginia (Chuck Robb), the mayor of Washington, DC (Marion Barry), and the administrator of the EPA (William Ruckelshaus) and convinced them to gather at a historic conference to sign the first Chesapeake Bay Agreement. The agreement established a top-level committee called the Chesapeake Executive Council, made up of governors, the mayor of Washington, DC, and the administrator of the EPA, which was to meet regularly and establish broad voluntary plans for reducing pollution. The agreement proclaimed that "a cooperative approach is needed . . . to fully address the extent, complexity, and sources of pollutants entering the Bay." The compact established an EPA office in Annapolis (the Chesapeake Bay Program office), which collaborated with other federal and state agencies to provide scientific data and advice (but not regulations or mandates) to the bay region states.

To feed the fires of public support for the campaign, Hughes gave speeches across Maryland and declared 1984 "the year of the Bay." "For three centuries, generation after generation has taken from the Bay. Let those who come after us be able to say that ours was the first generation to put something back!" Hughes said in his 1984 address to the Maryland General Assembly.

From 1984 to 1987, Hughes and the Maryland General Assembly approved more than $150 million in funding for the bay and created more than 270 state jobs dedicated to restoration of the estuary. Hughes signed legislation, sponsored by state senator Gerald Winegrad, a Democrat from Annapolis, that banned phosphates in laundry detergent, which had been polluting the bay with a nutrient that leads to excessive growth of algae. The legislation was controversial and opposed by Hughes's political polar opposite—the hot-headed, business-minded Baltimore mayor William Donald Schaefer, who sided with Baltimore soap manufacturers that didn't want to be regulated. Hughes also fought a successful battle to approve waterfront protection legislation called the Critical Areas Law, which limits development of waterfront areas within 1,000 feet of bay tributaries. "The net effect of that single piece of legislation was, in essence, to downzone about 10 percent of the landmass of the state," Hughes later remembered. "As you can imagine, this created a huge fight in the General Assembly. Yet, Maryland citizens, and eventually, Maryland legislators seemed to understand the purpose. . . . Without this law, we would have houses lined up along the shoreline of the Bay and its rivers like houses on a Monopoly board."

In 1985, Governor Hughes approved a moratorium on catching rockfish (also known as striped bass). Biologists at the Maryland Department of Natural Resources deemed a ban necessary after commercial landings of the East's most popular sport fish plummeted from 14.7 million pounds in 1973 to 1.7 million pounds by 1983, with recreational anglers experiencing a similar decline. Scientists worried that continued overfishing would cause irreparable damage to the species. In the summer of 1984, Maryland's secretary of natural resources, Torrey Brown, and his deputy, John Griffin, recommended an outright ban on catching rockfish, which would be unprecedented. "If that's what we've got to do, let's go ahead and do it," Hughes recalled telling Brown and Griffin. "It was that simple: One meeting, and we did it."

Actually, it wasn't *quite* that simple. Because striped bass are a migratory species that move up and down the East Coast, similar restrictions would be needed in all eastern states to make the moratorium effective. So Hughes had to ask Congress to support a proposal by a federal regulatory body, the Atlantic States Marine Fisheries Commission, which would then have to propose a similar moratorium on states from Maine to Florida. Congress passed a law called the Atlantic Striped Bass Conservation Act which granted the commission the authority to threaten federal fishing bans in other states if they did not voluntarily restrict their catch. This federal muscle was badly needed because Virginia did not want to cooperate. Maryland's ban lasted from 1985 to 1990, and in the meantime Congress worked to force all the other states along the East Coast—including the reluctant Virginia—to follow Maryland in cracking down on fishing. The protections up and down the East Coast proved spectacularly successful in bringing back the rockfish, with populations multiplying more than 10-fold in the decade following the moratorium, before leveling off.[5]

Hughes also took other actions to stop overfishing, which was a major problem. For the first time in state history, the Hughes administration in the early 1980s started requiring saltwater sport fishermen to obtain state licenses, with the money directed back into bay restoration efforts. "That caused howls of protest at first, especially from some of the conservatives from the Eastern Shore," Hughes wrote in his autobiography. "Fishing on the Bay, many thought, was a God-given right. But we thought it was high time to balance the ledger by imposing a user's fee."

Hughes also created the Chesapeake Bay Trust, a program that allows people to donate to environmental education and tree-planting projects by checking a box on their Maryland income tax forms or by purchasing "Treasure the Chesapeake" license plates. In addition, the Hughes administration approved a separate program to pay farmers to take steps to reduce the runoff of fertilizer from their fields, such as by planting strips of trees along streams to act as green filters.

Despite his earnest efforts and an outstanding record on environmental issues which changed the course of Maryland history, Hughes's political career did not end well. Term limits forced him to leave the governor's office in 1987, after eight years. Although he had originally campaigned as an anti-corruption candidate, he left with low approval ratings because of public anger over an unrelated scandal: the national savings and loan crisis. Hughes thought that it was unfair that his reputation was tarred by this fiasco because he had nothing to do with the banking meltdown. He even worked hard with the General Assembly to pass legislation to help protect people's savings accounts. But voters felt a generalized anger at politicians and had a vague sense of continuing corruption from earlier administrations. The damage had been done. When Hughes ran for an open seat in the US Senate in 1986, he lost to US Representative Barbara Mikulski, a once-obscure Baltimore City councilwoman, who went on to serve 30 years in the Senate, the longest of any woman in history.

Thinking back as he gazed out the windows of his home at the frozen waters of the Choptank River on a winter morning recently, Hughes reflected that his goal to quickly and dramatically "save the bay" was overly optimistic. He and other leaders did not realize the extreme complexity of the issue and the power of stubbornly entrenched financial and political interests. "I was telling people when we started the program, 'Don't expect anything overnight. It will take at least 10 years.' Well, that was the understatement of the year. It was a lot more difficult than we had anticipated. So it's taken a lot longer than we anticipated."

In his opinion, the problem has not been a lack of coordination between the federal, state, and local government entities involved in the bay restoration effort. Far more challenging has been convincing private property owners and businesses to sacrifice their own income or convenience for the common good.

> The most difficult part has been the people involved—the watermen, the crabbers, the farmers and others—and getting their support, because they've been doing things a certain way for a long time, and they don't like to change. A particular problem on the Eastern Shore has been chicken manure and how that is handled on farms. That has been a very difficult issue. It has been easy for poultry farmers over the years to just keep putting more chicken manure on farmland. But you can't continue to do that. The land only takes so much, and then the phosphorus [in the manure] runs off into the rivers and the bay. It's a big problem. But to get people to change the way they've been doing things for a couple of hundred years is not the easiest thing in the world, and certainly not the quickest thing in the world. It takes time, as we've seen. But progress is being made, and I think we'll get there eventually.

A major change that is still needed, Hughes said, is a transformation of personal attitudes by the vast majority of people, with a greater awareness of what they must individually contribute to make a cleaner bay. "I don't think the average citizen has had to change much about his or her way of life. The watermen—the people making their living off the water—have had to change and will have to continue to change. But the average people, the Bay Program hasn't really affected them, at least that I can see."

Bedeviling complications in the bay restoration effort arose in the decades following Hughes's tenure in office. After striped bass populations bounced back, scientists discovered that many of the fish were suffering from a mysterious bacterial wasting disease called mycobacteriosis. Hughes's acclaimed Critical Areas Law ended up having only limited impact in protecting waterfront land because—to win political support—the Maryland General Assembly included a "growth allocation" loophole in the law which allows local governments to choose a percentage of their land on which to still permit the giant waterfront developments that the law was designed to prevent.

Most importantly, the Chesapeake Bay Program itself turned out to be powerless and ineffectual. Supporters of the program argue that the voluntary and state-led "collaborative" nature of the program was politically necessary to secure the approval of Virginia (which had a history of strongly resisting mandates from Washington) and Pennsylvania (a politically influential swing state that owns no land on the Chesapeake). Unfortunately, however, the voluntary structure of the Chesapeake Bay Executive Council, as well as the purely advisory nature of the EPA Chesapeake Bay Program, meant that there was no centralized authority with enforcement power to compel the regional states to meet their pledged goals. And so the states made a little progress—especially by upgrading some sewage treatment plants—but fell far short of their targets. Looking back, critics point to the lack of teeth in the 1983 bay cleanup agreement, as well as in the follow-up agreements in 1987 and 2000, as an indication that they were naive from the start. The bay has "nonbinding agreements instead of enforceable laws . . . a bureaucracy that lacks regulatory powers and a severely impaired ecosystem that shows no signs of systemic improvement," wrote Howard Ernst, associate professor of political science at the United States Naval Academy, in his 2010 book *Fight for the Bay.*[6]

Hughes disagrees that the cleanup effort would have been more effective if it had been directed with a strong hand by the EPA-from the beginning. "I'm not sure of that at all. Top-down from the federal government might have caused more resistance," Hughes said. "I don't think that would have worked better, I really don't. I think things are more accepted by people if they know their states are doing it, and that it isn't being dictated by Washington."

He is optimistic that the most recent bay cleanup plan will work. It was established in 2010, with the EPA setting pollution limits for the states and threatening to penalize those that don't meet their targets. "Some targets have already slipped," Hughes admitted. "But yes, I think we'll get there. It's going to be difficult, but there are already signs of progress."

When Hughes was on the mound leading the Chesapeake Bay cleanup, he threw several very strong innings. Now he's on the edge of his seat in the stands, praying that the bay will make a big comeback.

PARRIS GLENDENING

The Green Governor and the Cell from Hell

P ARRIS GLENDENING eventually got somewhere in life. He earned a
PhD, became a professor, and then was elected the fifty-ninth governor of Maryland. But he wasn't born in an ivory tower or mansion. He
was raised in abject poverty in the Bronx, one of six kids of a gas station
owner whose business failed. When Parris was five years old, his family
piled into an Army surplus truck with everything they owned and drove
south in a desperate quest to scavenge a new life. In North Carolina, their
truck flipped and rolled down a hill. Nobody died, but their only asset—
their truck—was wrecked.

"We finally got to Florida courtesy of Travelers Aid [a charitable
organization that helps stranded and homeless people]," Glendening recalled at the age of seventy-two.[1] "But when we arrived, not only did we
not have any money, we did not have any personal possessions. Fortu-
nately, my mother's parents had a small farm there in Florida, and so we
stayed with them for a while."

Growing up in Hialeah and West Hollywood, Florida, Glendening
worked at a drug store in high school. He loved to go fishing in the Everglades with his dad. The younger Glendening aspired to go to college,
but his mother did not encourage education. So he paid his own way, first
attending a community college and then transferring to Florida State
University, where he earned a PhD in political science at the age of twenty-
five. Unlike some other doctoral candidates, however, Glendening also
had to labor in a machine shop to earn enough money to eat. And so, on
holidays and long weekends during his higher education, he would drive
an old Chevy eight hours back and forth between Florida State, in Talla-

hassee, and Hialeah, located west of Miami, so he could get paid to get sprayed with grease while shoving steel bars through a lathe.

During his long drives along the fringes of the Everglades in the early 1960s, Glendening saw the wetlands where he had spent the best times of his life fishing with his father: swamps alive with manatee, alligators, fish, and birds. He was stunned to see that the wetlands were being drained and paved for suburban subdivisions. The sight of the bulldozers kicked him in the gut. "It was a fundamental awakening to me—a deep sense of, 'This is wrong,'" Glendening said. "Part of the problem was there was no debate going on about suburban sprawl at the time. No one was saying, 'We should be preserving this land as open space.' That's just the way development was done back then."

Decades later, after serving as a professor of political science at the University of Maryland and then executive of Prince George's County, Glendening made land preservation and "smart growth" a top priority while he was governor, from 1995 to 2003. He preserved 400,000 acres of forests and fields that filter water flowing into the Chesapeake Bay, including large tracts of threatened waterfront property along the Potomac River (Chapman's State Park) and the Patuxent River (the Parris N. Glendening Nature Preserve at Jug Bay). Glendening said he was motivated in part by his experience as a young man watching the destruction of the Everglades, but also by his time as Prince George's County executive, watching white flight from Washington, DC, in the 1970s and 1980s, which spawned hideously ugly strip malls and subdivisions. This scattered development pattern—driven in part by racism—was expensive to society on several levels. The sprawl demanded expensive new schools, roads, and sewer lines—while feeding abandonment of buildings, businesses, and hope in lower-income neighborhoods. Meanwhile, the destruction of open space and forests ripped out the Chesapeake's natural filters. "You would drive down the road, and daily you would see the forests being cut down," Glendening recalled of his time as a suburban county executive.

In the Maryland General Assembly, Glendening championed a bill that was to become his greatest achievement: the Smart Growth Priority Funding Areas Act of 1997. The law was intended to stop sprawling development by concentrating growth into existing cities, towns, and priority areas that already have sewers and schools. The legislation was supposed to push construction away from farm fields, which are often more attractive to developers because they are blank slates that make it cheaper to stamp out cookie-cutter subdivisions. But Glendening—as a former county official himself—was politically cautious in designing the Smart Growth bill so that it gave full authority to the counties to designate these priority growth areas. The state's planning office could only comment on these local decisions, not block them—leaving all land-use decisions to county boards.

Despite the bill's deference to the power of local government, Smart Growth was fiercely opposed by the powerful Maryland Association of Counties and its allies in the General Assembly. The counties did not want to cede any local powers at all to the state—and they boasted an unholy alliance of financial interests backing them up. It wasn't just the county governments against Smart Growth; it was the development industry, bankers, and even the Farm Bureau, which opposes this model because it doesn't want anything to restrict the ability of farmers to sell their land and make money.

After a ferocious battle in the legislature and many compromises, the Maryland General Assembly passed the Smart Growth Act. It won national praise by environmentalists and editorial writers in the *New York Times*, the *Washington Post*, and elsewhere. The World Wildlife Fund touted the "Smart Growth Act" as "a gift to the earth," and Harvard University's Kennedy School of Government honored it with an "Innovation in American Government" award.

But despite the high accolades, the reality is that Glendening's Smart Growth law did not work to stop—or even slow—sprawl. This harsh judgment comes from scholars at the University of Maryland National Center for Smart Growth, whose creation Glendening supported. Sprawl actually *accelerated* after the law, according to the National Center for Smart Growth. The acres of land consumed for single-family homes outside of the "smart growth" areas (called "priority funding areas") increased from 75 percent in 1997 to 77 percent in 2004.[2] With regard to these "priority funding areas," "there is little evidence that after 10 years they have had any effect on development patterns," wrote Gerrit-Jan Knaap, the current director of the National Center for Smart Growth, in a 2009 article in the *Journal of the American Planning Association.* "Ten years after their official designation, priority funding areas are not well integrated in land-use decision making processes in many local jurisdictions. . . . These trends are going in the wrong direction."

The problem was that the "Smart Growth" idea sounded good, but the law itself was toothless. Leaving decisions about development entirely in the hands of county governments left the authority with officials who often viewed what others called "sprawl" as their main strategy for bringing in needed revenue for local businesses. Under the law, governors could theoretically decide to withhold state funding for roads and schools outside of areas the state deemed most suitable. But in reality, the state—including Glendening himself—rarely exercised this power, especially with politically influential suburban counties. And the relatively small amount of money the state could withhold was overwhelmed by the hundreds of millions of dollars the developers could throw around to open the doors for whatever they wanted to build.

Glendening—who is now president of a nonprofit called the Smart Growth America Leadership Institute—disputes some of the negative conclusions. He argues that "smart growth" was valuable because it helped to change the mind-set of many state and county officials so that they put a higher emphasis on land preservation. "If I could have a magic wand, I would have implemented a law with state control and true boundaries for growth vs. no growth, similar to those that were passed in Oregon and Washington State," Glendening reflected. But he added, "The truth of the matter is that strict boundary controls are not going to pass on the East Coast and in much of America today. You combine that with property rights, and with the Maryland tradition of local governance, and the idea would have been dead on arrival. And so we tried to figure out a way we could approach that idea [of strict boundaries for growth] and still be effective." In other words, Glendening—with his PhD in political science—made a political calculation and decided not to pursue the kind of policy he knew would be most effective. Even though he built much of his life and public identity around being a sprawl fighter, he gave up without even swinging for the knockout.

Former state senator Gerald Winegrad, a Democrat from Annapolis who headed the senate's Environment Subcommittee and is a friend of Glendening, commented, "Governor Glendening did the best he could, politically, in terms of Smart Growth. But did it work? No. Smart Growth is a total failure. In fact, sprawl has gotten worse."[3]

What happened with Smart Growth is a common political phenomenon that has hamstrung bay restoration efforts: many of the landmark environmental laws that lawmakers and advocacy groups (for their own political and fund-raising reasons) celebrate to voters and donors as major breakthroughs are, at the last minute, quietly riddled with so many loopholes and compromises that they produce, in reality, only marginal improvements, if any.

Glendening did succeed, however, in a different approach to conservation: he bought and preserved hundreds of thousands of acres of threatened rural land on the Eastern Shore, in western Maryland, and elsewhere, using state real estate transfer taxes set aside by a 1969 law called Program Open Space. And in 1998 he launched a new initiative called Rural Legacy, which uses state money to buy preservation easements on farms and other rural land, so that the owners can continue using the property if they promise that they and all following owners will not develop their land.

In another area of environmental policy important for the Chesapeake Bay, Glendening also hit rough political waters in 1997 when the bay experienced an ecological nightmare: an outbreak of a mysterious toxic microorganism called *Pfiesteria.*

That summer, more than 30,000 fish died on the Pocomoke River. Stranger yet, more than three dozen watermen and state wildlife managers who worked on the river reported memory loss, dizziness, and other related health problems. Scientists linked the illnesses and fish kills to manure runoff from the Eastern Shore's growing poultry industry, which seemed to be fueling the growth of a newly discovered single-celled organism called *Pfiesteria piscicida.*

The *Washington Post* proclaimed *Pfiesteria* the "cell from hell." And the exotic-sounding dinoflagellate became a national news story when Governor Glendening shut down the Pocomoke River to boating and fishing, because of the potential threat to public health. When more fish were found dead and with lesions, Glendening closed Kings Creek in Somerset County and a portion of the Chicamacomico River in Dorchester County.

"That was truly a crisis," Glendening reflected 18 years after the outbreak. "But there is an old saying that in every crisis, there is an opportunity. And all of a sudden, people understood, 'This is not just about the long-term health of the bay. This is not just about aesthetics or the environment. This is something that, had it affected our major rivers, we could have a disaster for our way of life.' That's when I think we hit a tipping point."

The *Pfiesteria* outbreak *could have* been a real tipping point for the bay, and some environmentalists argue that it *should have been*—a moment that inspired profound change, much like when Ohio's Cuyahoga River caught on fire in 1969. The intense media coverage of that burning river—including in *Time* magazine and *National Geographic*—helped persuade Congress to pass the federal Clean Water Act, which reduced sewage in waterways across the United States.

But a turning point did not happen with the Chesapeake Bay after the *Pfiesteria* crisis. In the months after the media frenzy, a blue-ribbon advisory panel chaired by former governor Harry Hughes recommended that the state take "immediate" action to halt the overapplication of poultry manure to Eastern Shore farm fields that were already saturated with phosphorus, so that the nutrient would not run off to feed blooms of algae and *Pfiesteria*.[4]

The problem was that the leaders of the Maryland General Assembly were steadfast against government regulation of farms, which were regarded as inherently virtuous and untouchable. "It was just a very, very steep political hill to climb—particularly because of the strong resistance of the legislature," Glendening said of his efforts to regulate poultry manure. "Most of the major committees were controlled by people from the rural areas, who were strongly opposed to any efforts there."

The image of the all-American family farmer—then and now—clashed with the reality that agriculture is increasingly dominated by large companies, excessive doses of fertilizers and pesticides, and factory-like warehouses packed with animals.

Busloads of farmers converged on the statehouse in February 1998 to loudly protest Glendening's proposal to regulate the application of poultry manure. Leading an intense fight against the bill backed by the governor was state delegate Ronald A. Guns, a fellow Democrat from Cecil County on the upper bay, chairman of the House Environmental Matters Committee, and an ardent defender of farmers, hunters, and watermen. "Agriculture did not bring this bug here," Guns said of *Pfiesteria*.[5] "Mother Nature did."

The powerful president of the Maryland Senate, Thomas V. "Mike" Miller, a Democrat from Prince George's County, was also among those who at first fought Glendening's push to regulate farms. "We're not going to be part of any bureaucratic rules or regulations that regulate the farmers off their property by mandating controls they can't accomplish," Miller said.[6]

After a fierce fight, Glendening accepted major compromises that severely weakened his bill. But the compromises worked, politically: in

April 1998, the General Assembly passed the Water Quality Improvement Act. The law was touted as the answer to agricultural runoff, the single largest source of pollution in the Chesapeake Bay. The law requires farms to write and follow plans that are supposed to control the application of manure and chemical fertilizer and reduce pollution in streams.

But the law did not work—if by "work" we mean improve water quality. Fifteen years later, water monitoring showed that phosphorus and nitrate pollution levels in the waterways of the Eastern Shore had not declined, and had even worsened in some places.[7] Although the fertilizer management plans required by the state in the early 2000s created some guidelines meant to discourage overapplication, the plans were often not followed by farmers.[8] A decade later, in 2012, farmers on the Eastern Shore were still applying *three times* more manure to their fields than their crops needed for their phosphorus nutrient content, and so the fields were still badly overloaded with phosphorus, which was running off into streams and the bay.[9] The 300 million chickens raised every year on the poultry farms on the state's Eastern Shore were producing 228,000 tons more manure than these farms could use for its phosphorus fertilizer content.[10] But the poultry growers still needed to get rid of their excess waste somehow, and they considered it a free form of fertilizer. So they kept dumping it in their fields in amounts far beyond what was healthy for the bay, in part because the manure also contains another important nutrient, nitrogen.

Reflecting on the lack of much improvement in the Chesapeake Bay since the *Pfiesteria* crisis, Glendening said part of the problem was weakness in the Water Quality Improvement Act itself, and part was a lack of its rigorous implementation later by state officials. Because of political compromises that lawmakers made with the farm lobby to get the bill passed, the legislation granted enforcement authority not to the Maryland Department of the Environment, but instead to the Maryland Department of Agriculture (MDA), whose primary responsibility is to promote the state's farm industry, not protect the bay. This created a conflict of interest that impedes bay cleanup to this day. "The problem is that the Department of Agriculture is trying to regulate its own team, its own group—the farmers," Glendening said.

As a result of another political compromise with rural legislators, the Water Quality Improvement Act required that farm nutrient management plans be kept secret, so that lawmakers, journalists, and the public cannot see how weak or strong the plans are (and whether or not the MDA is doing its job). The MDA succeeded in making 99 percent of farms write or obtain nutrient management plans, and the agency says it inspects more than 10 percent of these plans every year.[11] But this is largely a paperwork exercise. The state inspectors look at the documents but do not take samples from the fields or nearby streams to see whether the farms are, in fact, polluting the waterways or doing what they say they are doing. The plans themselves are fairly weak, in that they are designed primarily to maximize crop yield for the farmers, with environmental protection as a secondary goal. Moreover, many farmers openly admit they don't follow their plans. A 2015 study by Michelle R. Perez of the University of Maryland found that a majority (61%) of farmers interviewed on the Delmarva Peninsula said they don't obey the fertilizer limits in their plans.[12] "Several interviewed farmers, private planners, and fertilizer dealers stated they were actively evading the spirit and letter of the law because they kept double books (one plan to show an inspector and one plan to use to farm) or applied higher manure rates than they knew they should be using," Perez wrote. By contrast, only 17 percent of farmers (6 of 36) explicitly said they are "following the plan," an equal number (6 of 36) implied they are following their plans, and 5 percent (2 of 36) said they are "following soil tests."[13] Why such low compliance rates? First of all, because many farmers don't believe that the nutrient management plans are important. They reject the well-documented science that indicates that agriculture is the largest source of pollution in the bay, according to Perez's research. Moreover, many farmers don't want to produce only a modest amount of crops by using a modest amount of fertilizer. Understandably, they want to grow as much corn and other crops as they possibly can to make as much money as they possibly can, and adding more fertilizer helps to create a more robust yield, even though it is bad for the bay. Using poultry manure, in particular, is free or cheap for many farmers, so they want to use manure as fertilizer, despite the fact that it is often environmentally harmful because it is packed with excessive phosphorus content that runs off into streams. The bottom line is that what's good for the corn and the farmer's profit margins is different from what's good for the streams, the bay, and the general public.

"I think it was a lot of people's fault," Glendening said of the state's failure to reduce farm runoff in the decades after the *Pfiesteria* crisis. "But if I had to say it was anything, it was because there was a constant push by lawsuits [from farm groups fighting regulation by the state], by administrative delays, by legislative action to delay the full and effective implementation of the Water Quality Improvement Act. It was a constant battle from rural representatives from the shore, and from the big poultry industry, Perdue Farms, in particular."

Why the pushback from the industry? The farm lobby argues that more regulation of their industry is unnecessary and will put farmers out of business. They claim that other voluntary programs that use tax dollars to reward farmers for planting fertilizer-absorbing strips of trees along streams and planting cover crops—among other pollution control projects—have been successful in reducing runoff into the Chesapeake Bay. "Overall, the news is good," Valerie Connelly, executive director of the Maryland Farm Bureau, said in a July 2014 press release.[14] "We have consistently been ahead of schedule in Bay clean up. . . . Maryland farmers are environmental stewards." The industry and their allies point to computer modeling from the EPA Bay Program which shows that pollution from farms—in theory—*should* be going down, based on the good things farmers say they are doing.

In reality, however, water quality monitoring shows that phosphorus levels, often from animal manure, did not improve between 1998 and 2015.[15] "On the Eastern Shore the phosphorus levels have worsened over the past 10 years," according to written testimony that Scott Phillips, the US Geological Survey's Chesapeake Bay coordinator, submitted to the Maryland General Assembly.[16] "In the Choptank River watershed, for example [the largest river on the Eastern Shore], where the U.S. Geological Survey has the longest period of monitoring data, phosphorus concentrations have increased 48 percent over the last 10 years, and by 65 percent since 1985. The trends are worsening because phosphorus has accumulated in soils as a result of nutrient applications at levels in excess of what crops can use to grow."

The *Pfiesteria* crisis—and the water quality law passed because of it—should have inspired a reversal in this negative trend. But *Pfiesteria* did not light a fire like the one on the Cuyahoga River. This might have been because the "cell from hell" was much more ambiguous than a flaming river. In the years after the 1997 fish kills in the Pocomoke River and other bay tributaries, researchers questioned whether it was really *Pfiesteria* that had caused the fish kills, or, more likely, a different microorganism,

Karlodinium veneficum (which might also have multiplied because of high pollution levels). Looking back, *Pfiesteria* turned out to be not the "cell from hell"—and not even a big deal, because, as scientists later learned, *Pfiesteria* is common throughout the bay, but it rarely causes fish kills and has not caused any reported human illnesses since 1997.[17]

Glendening said that, in retrospect, he doesn't really know whether *Pfiesteria* was to blame. "Whether it is exactly this or exactly that, I don't know. But I know what the trigger was. And the trigger was the overloading of waterways from manure."

Almost two decades after the *Pfiesteria* frenzy, in January 2015, Republican governor Larry Hogan—under pressure from Democrats in the Maryland General Assembly—finally acted to impose regulations (called the "phosphorus management tool") that limit the overapplication of poultry manure. But these rules have a gradual phase-in over almost a decade, with a deadline of 2024.

So why did it take Maryland nearly a quarter century to address a problem—the overapplication of poultry manure to farm fields—that has been obvious since 1997? Glendening admits that he could have—and perhaps should have—ordered his own Maryland Department of Agriculture to impose rules to control poultry pollution years earlier. He said he did not push harder because of intense opposition from rural lawmakers. And he added that he did not want to undermine political support he needed to make progress on other priority issues beyond the environment, like improving funding for education. "The question is: Do you move 75 percent of your energy back to solving this problem [agricultural pollution in the bay]? Your personal, political capital? And the answer is, we did not. . . . But I believe, increasingly, that this issue must be addressed."

During his time in office, Glendening did attempt to do other things to control poultry pollution in the Chesapeake Bay. At the end of his second term, he issued permitting regulations that would have held the poultry companies, like Perdue Farms Inc., that own the chickens and make most of the profits responsible for the safe disposal of the manure owned by the contract growers. But this so-called co-permitting rule was challenged in court by the poultry industry and then abandoned by Republican governor Robert Ehrlich shortly after he took office in January 2003. Ehrlich also disbanded Glendening's Office of Smart Growth.

The biggest legacy that Glendening left, however, was in convincing the governors of Virginia and Pennsylvania and the mayor of Washington, DC, to sign an ambitious new bay cleanup agreement in June 2000. The agreement was Glendening's swan song, signed about two years before the end of his second four-year term. The "Chesapeake 2000" agreement was deemed necessary because the 1987 Chesapeake Bay agreement expired with its pollution reduction goals unmet.

Glendening's Chesapeake 2000 agreement was far more ambitious than the 1987 bay agreement. Few of the new targets, however, were ever achieved. Why? Because the states did not impose regulations to back up their aspirations and instead placed faith in voluntary initiatives.

In retrospect, Glendening—as an outgoing governor, with nothing to lose—had made a promise to clean up the bay by 2010 which he would never have to follow through on, and that his successors in Annapolis and other state capitals were not willing to keep. "I think it was a reach— maybe a bridge too far," he reflected in 2015.

Were the goals set by Glendening *impossible?* In 2004, a committee that reported to the bay region governors, the Chesapeake Bay Watershed Blue Ribbon Finance Panel, concluded that $15 billion would be needed to effectively launch a bay cleanup plan. But the panel concluded that the investment would be more than worth it, because the bay itself is worth more than a *trillion dollars* because of its tourism, boating, and fishing industries, as well as enhanced real estate values. A $15 billion cost of cleanup would have broken down to about $7 per person per month over 10 years for everyone in the Chesapeake Bay watershed.

"Maryland could have done more, I absolutely agree with that," said Glendening. "But the problem is, it's not just Maryland's responsibility. Also impacting the bay is Virginia, and in a major way, Pennsylvania, and even, because it's the headwaters of the Susquehanna River, New York. And the challenge was, during much of this period [when Glendening was in office] we had, particularly in Virginia, two governors—George Allen and Jim Gilmore [both Republicans]—who were not supportive at all. They were hostile. To put it mildly, they were not environmentalists and did not seem to have any major focus on the bay at all."

And Virginia and the other bay region states, beyond Maryland, matter a lot: about 71 percent of the nitrogen pollution fouling the bay pours from two states, Pennsylvania (44%) and Virginia (27%), while Maryland only contributes about 20 percent, and all the other jurisdictions (NY, WV, DE, and DC) combined are only responsible for about 9 percent, according to EPA data.

Because of this reluctance by some other states to take the pollution problem seriously, the answer, according to Glendening, is that the bay needs the federal government to assume even stronger control of the multistate cleanup effort. Glendening said that the current bay cleanup structure is still far too weak. In 2010, the EPA imposed pollution limits or a "pollution diet" for the Chesapeake Bay region states. But the system (also known as the Chesapeake Bay Total Maximum Daily Load, or TMDL) is still dominated by the states and only lightly guided by the EPA.

"Regardless of where you are ideologically about the role of the national government versus state and local government, there are some areas that absolutely require federal intervention," Glendening said. "And certainly one would be for the protection of civil rights. But another one is the environment. States cannot do this by themselves. And if we don't have more active federal intervention, we are never going to really succeed with the bay."

JOHN GRIFFIN
Watching Over the Wild

J OHN GRIFFIN WORKED for more than three decades as Maryland's secretary or deputy secretary of natural resources, dedicating much of his life to restoring the Chesapeake Bay. But, oddly enough, his inspiration grew from an alien landscape. The son of an Air Force pilot, Griffin grew up near a military base in Albuquerque, New Mexico. He remembers riding a train across New Mexico, gazing out the window at the panorama where he had collected the brightest memories of his life—hunting, camping, hiking, and fishing with his family. "The Santa Fe train had a big dome car on the back, and I remember watching the sun fall, and dusk spread out on all the plateaus and mesas, with all the different hues and colors," Griffin, 70, recalled at his home in Annapolis.[1] "At sunrise, when I woke up, I saw the same thing. Those landscapes branded or imprinted me like a fish or a goose. Something happens to me when I return there to New Mexico. It's where my soul is."

On many of his fishing trips with his father, they would stop and visit reservations, providing him with an early appreciation for Native American culture and beliefs. One of the Native American sayings he took to heart and frequently shared later in life was, "The way we treat the Earth is reflected in how we treat its inhabitants. And the way we treat one another is reflected in how we treat the Earth," Griffin said. "The idea brings together what I think is a very fundamental concept, and that is that human beings and human systems are part of a larger natural system. The planet is our life support system, and if we don't treat it well, we as a species are not going to be around long."

Eventually, Griffin went to college in Upstate New York. He thought he'd go into social work, but then he got mixed up in local government in

Maryland when a friend recruited him to work on the staff of the Prince George's County Council. At that job, Griffin met County Councilman Samuel Bogley, who in 1979 became lieutenant governor under Governor Harry Hughes. Griffin worked for Hughes as his environmental policy staff person, and then from 1984 to 1995 as deputy secretary of natural resources under Secretary Torrey Brown for Governors Hughes and William Donald Schaefer. Griffin was secretary of natural resources from 1995 to 1999 under Governor Parris Glendening and from 2007 to 2013 under Governor Martin O'Malley.

Before retiring from state government, he was Governor O'Malley's chief of staff from 2013 to 2015, a job that sometimes required 70-hour workweeks, including weekends. "You know, he would email me and others at midnight, at 3 in the morning, on his Blackberry when he couldn't sleep," Griffin said about Governor O'Malley. "He would get on the Bay-Stat website [which lists Chesapeake Bay pollution data], and he'd say, 'I don't understand? Why isn't this moving up? Or why is this doing this?' ... I was like, 'How do you have the time to do these things?'"

Griffin changed the culture of the Maryland Department of Natural Resources (DNR). For too long, the fisheries management agency had an overly cozy relationship with watermen and the seafood industry. Griffin put scientists instead of political hacks in charge and focused on conservation. Managers who worked under him said he gave them confidence and raised morale, because they no longer had to look over their shoulders, worrying that a state delegate would get them fired for arresting the wrong waterman for oyster poaching.

"I don't think there is anybody who has done more to support and advance bay restoration than John Griffin," said Eric Schwaab, deputy secretary of natural resources under Griffin. "He was able to work aggressively to set the right tone and the right pathway within the agency, from a leadership perspective. But he has also developed strong partnerships with other organizations and agencies that needed to be aligned toward bay restoration. And he's been able to help chief executives—governors and on down—see a pathway forward and execute a pathway forward in bay restoration."

As over-secretary of Governor O'Malley's Chesapeake Bay Cabinet from 2007 to 2013, Griffin led Maryland's efforts to create a plan to meet new pollution limits for the nation's largest estuary, set by the EPA in 2010.

"John understands policymaking, he understands natural systems, and he understands what motivates people," said Ann Swanson, executive director of the Chesapeake Bay Commission. "And so John was able to create programs that really transformed how we manage not only the Chesapeake Bay, but also the state's fisheries and parks. John is one of the true great visionaries of this century. He changed Maryland for the better."

Griffin refuses to accept credit for any of this, consistently deferring to the governors he served, and heaping praise on the biologists and other professionals inside and outside DNR who advised him. But those involved know that it was Griffin who was really the driving force behind many of the policies that brought some life back to a dying bay.

For example, early in his career, after a crash in striped bass populations, Griffin and his boss, Maryland secretary of natural resources Torrey Brown, convinced Governor Harry Hughes to impose a moratorium on catching striped bass from 1985 to 1990. "When we proposed that, it was a major controversy. I mean a major controversy," Griffin recalled. "We had public hearings on the regulations, and we were getting death threats. The threats came in phone calls and in the mail, saying 'I'm going to kill you, you expletive deleted.' We had to have police at the public hearings."

Hughes pushed ahead with the moratorium, anyway. And afterward, Griffin was grateful to see striped bass—Maryland's state fish—recover in a dramatic way. But he was also traumatized, because he genuinely liked the watermen who were seething with rage over the issue. The intensely negative experience made Griffin gun-shy about imposing any future bans on fishing, even when necessary. Griffin, who grew up loving to fish and hunt, came to see even temporary bans on fishing and hunting as blunt tools that reflect badly on the agencies that impose them. A wiser fisheries management agency, Griffin concluded, would take moderate measures well in advance of problems to prevent crashes in populations, so draconian action later would not be necessary.

That philosophy informed his later, more limited actions to reduce the catch of oysters and blue crabs in the bay. It is possible that because of Griffin's increased caution as a regulator and compassion for people, these nonhuman species—which may have benefitted from temporary bans on harvest—did not get the tough-minded protections they really needed (and that even watermen may need, in the long run). It is hard to know, however, because striped bass, oysters, and crabs all have very different reproductive patterns, life cycles, and sometimes weather-dependent booms and busts.

In 2007, the Chesapeake Bay's blue crab harvest hit a record low. Biologists concluded that the collapse was caused in large part by overharvesting by watermen, who routinely caught more than half of all blue crabs in the bay every year, and sometimes as much as 79 percent.[2] In an effort to rebuild populations of female crabs to sustainable levels, Griffin and O'Malley in 2008 imposed restrictions (but not a ban) on the commercial catching of female blue crabs. To help compensate watermen for their losses, Maryland paid them to remove abandoned crab pots from the bay. And Senator Barbara Mikulski led an effort to have the blue crab fishery declared a national "disaster," providing millions of dollars in federal funds to the industry to help offset the losses from the harvest restrictions.[3]

Virginia, led by Governor Tim Kaine, a Democrat, joined with Maryland in a coordinated plan to protect blue crabs. It was a rare show of teamwork between states that had a history of conflict over the management of oysters, with the most dramatic example being the "oyster wars" between watermen of the rival states which erupted into gunfire in the late nineteenth century. In 2008, Virginia banned a longtime waterman practice that was devastating to blue crab populations: dredging for pregnant females in the winter as they hibernated in sediment on the bottom of the southern bay. As a result of the ban on the winter dredge, blue crab populations nearly tripled between 2008 and 2012.[4] "Obviously, we were very pleased with the result, as was Virginia," Griffin recalled.

Unfortunately, after that four-year surge in blue crabs, their numbers fell off sharply again in 2013, perhaps because of bad weather conditions— including a cold snap, unfavorable winds and currents, and unusual temperature and storm conditions that may have brought a large number of predators (specifically, crab-loving red drum fish) into the bay. As a result, by 2014, the blue crab numbers had fallen nearly as low as they had been in 2007, although they improved slightly again in 2015 and then surged in 2016 in an erratic up-and-down pattern that is common for crab populations.[5]

Beyond the management of crabs and striped bass, Griffin's agency also cracked down on illegal fishing for striped bass in the bay—a chronic and major problem. For example, on January 31, 2011, natural resources police confiscated more than 10 tons of striped bass from four illegally anchored gill nets near Bloody Point Light, south of Kent Island, in the Chesapeake Bay.[6] To deter incidents like this, the Department of Natural Resources began deploying radar and camera systems to keep watermen out of no-harvest zones, and it obtained court orders to attach electronic tracking devices onto the boats of repeat offenders. To convince the courts to take poaching cases more seriously, Griffin's agency worked with the court system to establish special "natural resources" days in the courts once a month, when all of the fishing and hunting violations could be handled by a single prosecutor.

Perhaps Griffin's most important legacy, however, was turning around Maryland's failed oyster restoration program. Oyster populations in the bay had plummeted to less than 1 percent of historic levels by the 1990s, and they kept falling despite millions of taxpayer dollars being invested to plant young oysters atop beds of transplanted oyster shells. Watermen would just scoop the oysters right back out of the water again after they were planted. It was a wasteful, government-funded "put and take" fishery that gave the misleading impression of restoring the bay's oyster population as the shellfish kept declining.

"We had been doing an oyster repletion program in which we'd dredge up old historic oyster shell, clean it, and deposit it around the bay in areas where watermen would harvest it," Griffin said. "But it became—because of pollution, because of parasites, because of overharvesting—a pretty lame program. And the scientists concluded that we were also unwittingly perhaps spreading the oyster diseases Dermo and MSX around the bay by moving oyster shells. So we phased that program out."

Instead, under Griffin, Maryland created no-harvesting zones to protect 24 percent of the bay's remaining quality bottom habitat for oysters. And the state started planting millions of baby oysters in these sanctuaries. Maryland also offered low-interest loans to encourage watermen to abandon their traditional dredging up of wild oysters and instead switch to aquaculture. "It was challenging because the oystermen were used to being hunter-gatherers," Griffin said. "They were not used to doing all the work that's involved with farming oysters in floats, bags or cages on the bottom."

Many watermen are angry about the changes to their way of life and blame the state government. Robert T. Brown, president of the Maryland Watermen's Association, points the finger at Governor O'Malley for the changes led by Griffin. "Worst governor we ever had," Brown said of O'Malley. "Just the way he taxed us to death. The way he turned around the Department of Natural Resources was so hard on us, including with the oyster sanctuaries. The state took over the bottom of the bay."

Because of the changes in state policies in 2009, Maryland went from almost no oyster farming to 153 aquaculture businesses that held leases on 4,340 acres of bay bottom by 2015. In addition, wild oyster populations and harvests in the bay more than doubled between 2010 and 2014, although they remain at less than 1 percent of historic levels.[7]

Under the shift in policies that Griffin established, a growing percentage of oysters from the bay will be from aquaculture. This is an important step forward, because oyster farming is far more sustainable than wild harvest and does not rip up the reefs and grasses on the bay bottom, as oyster dredging does. Creating sanctuaries for less than one-quarter of the remaining oyster reefs was a fairly modest step. But after Griffin retired, watermen immediately started lobbying new governor Larry Hogan, a Republican, to eliminate or shrink the new sanctuaries.

Despite this backward pull, it is clear that Griffin deserves praise for moving oyster management in the right direction: toward treating the bay with respect and restraint, with the realization that human beings are part of a larger natural system.

BONNIE BICK

A Soft-Spoken Warrior for the Chesapeake's Forests

B ONNIE BICK is a seventy-three-year-old former flower child and preschool teacher with a gentle voice. She has little money and few possessions and loves walking in the woods near her small brick house in southern Maryland. But in her watery blue eyes is a ferocity that has stopped bulldozers. Over three decades of unpaid conservation work, Bick quietly orchestrated statewide opposition to a series of massive development projects and in so doing saved thousands of acres of forests in a state increasingly devoured by subdivisions and strip malls.

A side effect of her success, however, was payback. She infuriated local officials in Prince George's County by fighting a casino, condo, and shopping project called National Harbor. In 2004, local officials threatened to use eminent domain to demolish her home in Oxon Hill to build a stormwater pond as part of a road-widening project to serve the casino.

"It was a terrible situation to be in, because I really love that house," Bick recalled.[1] She had grown up in the two-story home on Oxon Hill Road, with a view of the Potomac River from its back porch. She was attached to the house in part because her father had designed it, and he was a smart guy—a rocket scientist for the US Naval Research Laboratory.

The timing and circumstances surrounding the condemnation of her home seemed suspicious.[2] First of all, the developer, Milton Peterson, had already told the county he didn't need the road widened. And then Bick hired an engineer who concluded that it would be impossible to build a stormwater pond on her land, because of its steep slope. More strange still, Bick's home was the only one identified by the county that "had to be" bulldozed for the road-widening project.

Many saw the threat as an attempt to intimidate Bick. It didn't work.

The county offered her $200,000, but she did not accept the check, saying that it wasn't about money. She kept fighting the casino, rallying her neighbors, protesting at county council meetings, and calling the *Washington Post* and anyone else who might help.

In the end, the road was not widened. Bick kept her home.

"If taking my house would have made the community better, I would have yielded to it without regret," Bick said as she strolled through her backyard. "I would have been unhappy to lose it, of course. But the bigger issue was the National Harbor project, which was to replace trees and greenspace along the Potomac with a giant retail and gambling complex and bring in casino gambling."

National Harbor was eventually built, but it included a public walkway along the Potomac which Bick and allies had pushed for, as a way to provide public open space beside the water. Several other major development projects that Bick opposed across the region were defeated.

Among land conservationists, Bick is a hero—a sweet but tenacious street fighter who outsmarts lawyers and outlasts megaprojects that most people thought were unstoppable. In the end, it was Bonnie who proved unstoppable. For example, in a different battle, she took on the entire political and business establishment in Charles County to kill a proposed highway called the Cross County Connector, which would have served at least 8,000 new homes in what would have been sprawling subdivisions. Bick and her allies stopped the highway in its tracks, and there is now a proposal for a bike path surrounded by trees.

"That was a very exciting end for a terrible, long ordeal," Bick reflected. "It was like a miracle that we won."

Former Representative Donna Edwards, a Democrat who served central and southern Maryland and who has known Bick for 18 years, said that Bick deserves tremendous credit not only for pushing back against poorly planned development but also for advocating for the health of urban communities. Bick, who is white, has warmly welcomed minorities into a conservation movement that is often very white, said Edwards, who is African American. Bick worked with Edwards and others to try to protect green space and local businesses in Oxon Hill, which is 75 percent black.

"Bonnie is a different kind of environmentalist," Edwards said. "Because not only does she believe in preserving the environment, but she also believes in making sure that vulnerable communities are protected as well. Not all environmentalists take that tactic, and that's why her work has been so important and so deep."

Developers and their allies often ridicule Bick and her kind as "NIMBYs" ("not-in-my-backyard" fanatics) or, worse, "BANANAs" ("Build Absolutely Nothing Anywhere Near Anything"). But this is not an accurate reflection of Bick's views or approach. For example, she strongly supports development in urban areas like Baltimore and Washington, DC, which have vacant land and abandoned homes. She is driven by a philosophy of improving cities so that rural areas can remain rural. But she also believes in saving wildlands as a home for nonhuman life. "I'm a big supporter of the protection of biodiversity," Bick said. "Large properties that haven't been disturbed are usually very biodiverse. I am a great believer in thinking globally and acting locally. If we could just get a lot of people motivated to protect the most valuable resources in their local areas—that would be good for everyone."

Hers is NIMBYism flipped from vice to virtue: the idea that each person has a responsibility to be vigilant, that each soul is a soldier in a secret army assigned to defend the remaining green around us. Nobody can save the world. But each person can save a few trees down the street, or work with neighbors to haul tires and trash from a local stream. If each person contributes a little, the combined power is enormous.

"It's not about the individual," Bick said. "It's working together and trusting other people. That's what brings success and meaning."

Bick was born on August 25, 1943, in Hackensack, New Jersey, one of three children of a physicist, Frank Ferguson, and his wife, Mary. She attributes her emotional sensitivity to her mother, who lost both of her parents to the influenza epidemic of 1918 when she was only nine. She had to raise her two younger sisters herself, even though she was also only a child. "I remember seeing my mother crying when she read the newspaper," Bick said. "She had a tough early life, so she could feel others' pain." The family moved to Maryland in the 1940s, when Bonnie was about three, so that Frank could take a job with the US Naval Research Laboratory, working on radar and submarines and then designing missiles. After the war, he assumed a leading role in engineering guidance and control systems for the Vanguard rocket program, which was an early part of the space race between the United States and the Soviet Union as they competed to launch satellites into orbit.

When Bonnie was six, the family moved from Forestville, Maryland, in Prince George's County, to Oxon Hill, where her father built the two-story red brick house on a hill over the Potomac River. "It was extremely rural at the time," Bick said. "As a little girl, I was allowed full access to the woods, and we were surrounded by woods."

After graduating from Oxon Hill High School, she went to the University of Miami and then transferred to the University of Maryland, College Park. She graduated in 1965, moved to New York City, and worked at Bloomingdale's as an assistant clothing buyer. But she didn't find the job fulfilling, so she quit and started working as a teacher in Brooklyn.

During the Vietnam War, she became very involved in anti-war protests and was arrested three times, including once outside the State Department in Washington, DC. She recalled that she and others were sitting on the sidewalk, holding anti-war signs and chanting, when a police officer shoved her to the ground and wrenched her arms into handcuffs. "It hurt, and so I looked at him and said, 'I am doing this because I love you, too.' So he immediately stopped hurting me . . . and said he was against the war, too."

Bick married, had a son, and raised him in Tarrytown, New York, about 25 miles north of New York City. She moved back to Maryland in 1985 when she separated from her husband. Eventually, she lived once again in the red brick house where she grew up.

She became interested in conservation in the late 1980s, when a friend asked her to attend a county planning board meeting. Local officials were discussing the construction of an immense "Outer Beltway" highway project around Washington, DC, which would have crossed the Potomac River with a new bridge near Bryans Road, about 15 miles south of where she lived.

"I understood that huge amounts of growth would be spurred by the Outer Beltway and that it would cause ecological havoc," Bick recalled. "I saw what had happened with the construction of the original Capital Beltway [from 1955 to 1964]. And I knew that another ring would be equally damaging, if not more so, to the Chesapeake Bay."

The Outer Beltway was eventually defeated. But speculators bought up land along its route, including on both sides of the Potomac, and proposed a series of mega housing developments. Bick and her allies spent decades fighting this sprawl. In the 1990s, Bick was a leader in a grueling but eventually successful multiyear war to stop a 4,600-home subdivision, office complex, and golf course called Chapman's Landing. The project was proposed on 2,250 acres on the banks of the Potomac, across from Mason Neck and about 6 miles south of Mount Vernon, George Washington's historic home.

"Not only would Chapman's Landing have destroyed the quality of life for the people in the area, but it would have also changed the nature of the lower Potomac," Bick said. "It would have put a whole new city right on the river. It was just a terrible threat."

In the end, Bick and others convinced Maryland governor Parris Glendening to halt the project in 1998. Glendening used $25 million in state funds, plus $3 million from a nonprofit foundation, to buy up most of the land and create Chapman State Park and Chapman Forest.[3] Now most of that land has been protected as a Maryland state wilderness area.

That victory inspired Bick and others to try to also protect a nearby stream that flows through the heart of the area: Mattawoman Creek. It is one of the most fertile fish breeding grounds in the Chesapeake Bay Watershed. Bick worked with allies to protect an additional 1,000 acres beside the creek which were threatened with future potential development. In 2011, Bick was part of an organization, Friends of Mattawoman Creek, that worked with allies to halt the Cross County Connector highway project. It would have crossed the creek's watershed and required the clearcutting of 74 acres of forests and the destruction of 7 acres of wetlands.[4] The Mattawoman decision—which hinged largely on state denial of a wetlands destruction permit—was a rare one. The Maryland Department of the Environment rarely turns down permits to build in wetlands, approving more than 99 percent of the applications for permits and authorizations in some years, according to state records.[5]

Bick cofounded the Mattawoman Watershed Society with Jim Long, a physicist. Long said that Bick possesses a combination of talents that allow her to work well in the coalitions needed to achieve any kind of environmental victories. "She's pretty unusual in that she can work at all levels. She will knock on doors, and she will call on governors," Long said. "She's fearless in approaching the upper echelons of government, and yet she is still very warm and can work at a grassroots level. And, most importantly, she has vision. Before anyone else, she recognizes when there is an opportunity or a problem."

Devoting a lifetime to fighting against the grain of modern America has been emotionally grueling work that has required great financial and personal sacrifice for Bick. Although she works around the clock for the Mattawoman Watershed Society and other volunteer organizations, she has virtually no income.

"Bonnie is a minimalist," said Congresswoman Edwards. "She doesn't believe in accumulating lots of things and she's not materialistic. So I think in her personal life she has minimized her needs, and so that actually has liberated her when it comes to volunteering and doing community work for nothing."

It hasn't been easy, however. Bick has been threatened with lawsuits and subjected to verbal abuse. And in 2006, she nearly died in a car accident driving away from a land preservation press conference.

"I was hit by another car and had nine broken ribs and my lungs collapsed," Bick remembered. "My pelvis was broken in six places and my sacrum was broken and my bladder was smashed and ripped up and my liver was lacerated. Everything inside my seat belt was crushed. The doctors put me into an induced coma."

After her brush with death, she woke up in her bed at Washington Hospital Center and almost immediately continued to meet and scheme with her fellow activists on ways to protect their beloved Mattawoman Creek from development. By keeping her posted on what needed to be done, her friends gave her a reason to recover from her horrific injuries and keep living. One of those who visited her in the hospital was fellow conservationist Linda Redding.

"It's amazing: there are literally thousands of acres across Maryland that have been saved because of Bonnie's efforts," Redding said. "A lot of people say, 'Oh, I can't make a difference. It doesn't matter.' But that's just wrong. Bonnie has shown everyone what one person can do."

MICHAEL BEER

The Lorax of Baltimore

A MID THE BLACKTOP of Baltimore is a forest with a stream running through it. Old oaks and tulip poplars form a shady vault over a creek with a path beside it. But it was not the city government that kept the Stony Run Park free of trash and tended like a garden.

It was the Lorax. On Saturday mornings, out in the park, residents of the Evergreen neighborhood would see an old man with big hands, planting trees and flowering bushes, picking up trash, and ripping out invasive vines with incredible energy.

His name was Michael Beer.

"He looked sort of like a miniature gnomish Santa Claus," recalled his friend, Marian Perry Tamburrino, a retired high school teacher.[1] "He would often wear this red felt hat, and he had a beard, and his eyes never ever stopped sparkling. And he just would smile at anything that was a challenge. He'd say, 'Well, let's see what we can do about that.'"

Beer was the cofounder of the Jones Falls Watershed Association, a nonprofit group that organized thousands of neighbors and students to clean up urban streams and plant native trees.

In a city where it is hard to get the government to do even small things, Beer used his optimism and his persistence to convince the city to shut down an entire interstate—the Jones Falls Expressway—once a year for his Jones Falls Festival. Rock bands played on the roadway, and thousands of people bicycled, Rollerbladed, and walked down the highway over the stream.

Beer also single-handedly convinced the city to release water from the dam at Lake Roland to allow kayaking and canoeing through the city on the Jones Falls during his festivals. The Stony Run flows into the Jones

Falls, which empties into Baltimore's Inner Harbor and eventually the Chesapeake Bay.

"I think the Stony Run is an example where one person can really make a difference based on passion and hard work and a little elbow grease," said Halle Van der Gaag, former executive director of Blue Water Baltimore, which merged with the Jones Falls Association in 2010. "This stream would not look like it looks now, and this park wouldn't look like it looks, without Dr. Beer's advocacy and his real commitment to woodland plants, native plants, and how you have to maintain and care for trees."

Dr. Michael Beer—with a PhD in physical chemistry—was not just an old guy who picked up trash in the park. Born in Hungary to a family that immigrated to avoid the Nazis, Beer was also a groundbreaking scientist of international fame, although you'd never know it, because he never talked about his accomplishments. He was a global pioneer in the development of the electron microscope. As a biophysics researcher at Johns Hopkins University, he and a colleague built one the world's first scanning electron microscopes capable of capturing images of individual atoms.

"Michael was very humble, and yet he had accomplished fascinating things," said Tamburrino. "He brought the electron microscope to Johns Hopkins. He also showed how the electron microscope worked to my group of high school students, without even letting them know that he had designed it and built it."

After he retired from Johns Hopkins in 1995, Beer did not retreat or sink into isolation or depression—even when hard times came, like when his wife, Margaret, died of cancer and then his daughter Wendy died. He reached out to his neighbors, inviting kids and adults to Charlie Chaplin movie nights at his home, serving popcorn and apple cider. Even when he moved to a retirement home, he campaigned successfully for the institution, Roland Park Place, to install a green roof.

They say you can't beat city hall. But if you are Michael Beer, you can charm it and improve the world around you.

Michael—my friend—died of a heart attack at the age of 88 on August 22, 2014. He was an inspiring example of how to age with meaning, love your neighbors, and live with nature.

There are more than 100,000 streams and rivers that—like the Stony Run—flow into the Chesapeake Bay. If each of them had someone like Michael to guard it, the bay would run clean and we'd all have his sparkle in our eyes.

CAROLE MORISON

Free as a Bird Now

CAROLE MORISON raised chickens for Perdue for 23 years on Maryland's Eastern Shore. Then one day, she decided she'd had enough. It was the day the company told her to block all the windows of her chicken houses, so that Perdue's 54,400 birds would live in constant darkness.

Why darkness? "Well, when it's dark inside the chicken houses, the chickens are docile," Morison explained.[1] "That way, they are not wasting energy, which means feed. They just plop down and convert that feed into pounds of meat."

Morison refused, and her contract was terminated. "It was 'either you do what we say, or you get no chickens,'" Morison said of working on contract for Perdue. "I became disenchanted with the whole system. There were so many chickens packed in there, and they were so heavy, they would not get up to move. To me, that's no way to raise an animal."

The Salisbury, Maryland–based Perdue, the third-largest poultry company in the United States, has a slightly different version of the breakup story. Julie De Young, a spokeswoman for Perdue, said the company required all their contract farmers, including Morison, to close off all chicken house windows—day and night—so that large fans at the end of the metal buildings could suck air through in what is called "tunnel ventilation." Perdue required that growers like Morison borrow money to build tunnel ventilation systems, although Perdue earns most of the profit from the systems. Carole Morison refused to borrow $150,000 to build the new ventilation system, and so her contract was terminated.[2]

"It's correct that we stopped working with Carole," De Young said. "As Perdue has continued to improve the housing environment for our birds, we do work with growers to continue to modernize and upgrade their poultry houses. . . . The Morisons—along with a handful of other growers—chose not to re-invest in their houses, and so we did not renew their contracts."

The divorce between Morison and Perdue was final in 2008. After three years of wondering what her future would hold, she had a revelation: she would go into business for herself. She purchased 630 Rhode Island Red hens, good for laying eggs. She cut holes in the sides of her two chicken houses to allow sunshine and ventilation. And she freed her birds to explore outside and eat bugs, worms, and grass, as well as the grain that she feeds them. "We are now a pasture-raised egg farm, proudly certified as animal-welfare approved," Morison said outside her remodeled chicken house, as she watched her birds graze and peck. "The chickens go in and out as they please, which is more enjoyable for them and us."

Morison believes that their new diet is healthier than the antibiotics, chicken parts, ground-up feathers, grain meal, and arsenic that Perdue forced her to feed to the birds.[3]

Most surprising, however, is that Morison—as an independent businesswoman—said she earned about 50 percent more money (about $26,707 annually) selling her eggs to Whole Foods than the $18,000 per year she netted raising chickens under contract for Purdue. Her eggs were proudly labeled with her own name on the cartons: "Bird's Eye View Farms— Carole Morison." Whole Foods sold Carole Morison's pasture-raised eggs—under her name—for $5.99 a dozen (about twice the price of supermarket eggs from industrial-scale farms).

"I'm a completely different person now—my attitude is not as bad as it used to be," Morison said, striding across her farm on a sunny afternoon. "I really enjoy the independence of being a business owner. It's more American. You take a lot of pride in what you are producing. And it's definitely a sustainable business. Whole Foods buys every single egg that we can produce and our business is continually growing. Back when we were working for Perdue, we all had to have other off-farm jobs full time, because we didn't make enough money. We had to go out and earn extra money just to run the chicken business. It was constant stress."

The poultry industry dominates Morison's community on the Eastern Shore, where Arthur Perdue in the 1920s invented the factory-style poultry production methods—with giant metal buildings packed with thousands of animals—that have now spread around the world and to other livestock industries, including the raising of hogs and turkeys.[4] More than a half billion chickens worth $3.2 billion are raised every year on the Delmarva Peninsula by the industry, which employs about 14,600 people.[5] A side effect of this system is a vast volume of waste that pollutes the Chesapeake Bay. Farmers like to spread poultry manure on their corn and soybean fields as a free (or cheap) form of fertilizer. But on Maryland's Eastern Shore, poultry farms generate far more manure—about half a billion pounds too much every year—than local farms can use for its nutrient (phosphorus) content. The excess phosphorus runs off to pollute streams and feed algal blooms, creating "dead zones" in the bay.[6] Despite claims by industry of progress in controlling this chronic problem, an analysis of water quality monitoring data from 2003 to 2013 in eight major waterways on Maryland's Eastern Shore found no improvement in runoff—in fact, worsening phosphorus pollution levels were noted in three rivers, the Nanticoke, the Sassafras, and the Transquaking.[7]

The industry is hard not only on waterways but also on its own contract farmers. The system is designed to maximize the profits and minimize the environmental liability for the poultry companies, while loading onto the farmers most of the risks, debt, and manure handling responsibilities. Companies like Perdue, Mountaire, and Tyson own the chickens and provide the feed for the birds. The contract farmers own the land and the waste produced by the chickens, and they supply the labor. The farmers must borrow hundreds of thousands of dollars to build chicken houses to the companies' exact specifications, and they must raise the birds precisely as dictated by the companies. If the farmers fail to follow orders, the companies can threaten to withhold the next batch of chicks—leaving the farmers stuck with huge debts and no way to meet their loan payments. According to a 2011 report by the University of Maryland Extension Service, "Potential disadvantages for the [contract farmers] include the elimination of extra profit opportunities, sharing or giving up some control of management decisions, and no equity in the birds."[8]

In a 2009 letter to US Attorney General Eric Holder, Morison explained the financial pressures this way: "We invested over a quarter-of-a-million dollars to get into the business. I know of other growers that invest over $1 million to get in. . . . We have had the contract termination threat used on us several times leaving us to worry constantly that our contract would be terminated if we did not do exactly as the company wanted us to do. . . . Basically, growers keep their heads down and their mouths shut or prepare their self for retaliatory actions."[9]

The farmers are also burdened with all of the environmental responsibilities, which may include obtaining state and federal permits and plans
intended to keep waste out of streams (although not all poultry operations get these permits or follow these plans). In return for all of these
burdens, the paychecks for the family farmers are not large, especially
compared to the debt they bear. For example, in 2011, it cost a contract
farmer about $318,900 to build a chicken house, about 60 feet wide and
550 feet long, to hold about 30,000 birds, meeting specifications required
by the poultry companies. The average income per year for a contract
farmer was about $17,931 per poultry house, according to the University
of Maryland Extension Service report.[10] (Morison says this estimate is
high—about twice the $9,000 per house net income she actually received,
when the costs of cleaning manure out of the houses and paying the electric bills and other expenses are taken into account.) As the poultry industry grows with consumer demand and more houses are built on the
Eastern Shore, some contract farmers these days are building as many as
a dozen poultry houses, each larger than in the past—some now twice the
length of a football field, to hold about 36,000 birds at a time per house.[11]
A dozen of these large poultry houses would cost a family farmer almost
$4 million to build, all of which must be borrowed by the farmer. "Chick-
ens reach slaughter and processing weight in about six or seven weeks,
but loans taken out to build henhouses can last for more than a decade,
making many chicken growers entirely dependent on a series of flock-to-
flock contracts to repay their debts," according to a 2015 report by Food
& Water Watch. "The poultry sector is less like a free market than abject
serfdom."[12] It is not a coincidence that the modern contract poultry growing system spread in the same geographic locations (including in the
Deep South and on Maryland's Eastern Shore) as the exploitative "share-
cropping" farming arrangements that took advantage of freed blacks after
the Civil War, according to a Johns Hopkins Bloomberg School of Public
Health scientist, Dr. Ellen Silbergeld, who published a book on the industry called *Chickenizing Farms and Food.*[13]

Morison didn't grow up on a farm, and she was shocked to find out how
little power farmers had over their own farms. Carole Nowakowski was
born and raised in the beach town of Rehoboth, Delaware, as the daughter
of a home builder and his wife, a gym teacher. At the age of twenty-nine,
after working as the manager of a bakery, Carole met and married a lifetime farmer, Frank Morison, and moved to his place on Byrd Road in

Pocomoke City. They had three children, Frank Jr., Christy, and Natalie. To make a little extra money, in addition to growing corn and soybeans, they built two chicken houses to raise almost 300,000 chickens a year for Perdue. Raising chickens paid so little, however, that Frank also took a job working as a parts manager for the John Deere dealership in Pocomoke. About four or five years into her poultry experience, in 1991, Carole said she became so disturbed by the power imbalance between the poultry companies and contract farmers that she began to reach out to other farmers to try to organize them to obtain better working conditions through a group called the Delmarva Poultry Justice Alliance, a project of the Episcopal Diocese of Delaware. "It's a feudal system," Morison said of growing chickens on contract.

According to a US Department of Agriculture study, between 2009 and 2011, almost one-third of poultry contract farmers across the United States were required by poultry companies to make capital upgrades like the one that Perdue tried to force on Morison. These upgrades required poultry farmers to borrow an additional $142,000 per farm on average, or a total of $637 million.[14] For most farmers, the debt burden for the chicken houses far exceeds what they owe on their own homes, which are typically used as collateral. This puts the families at risk of homelessness if something goes wrong. In a 2011 survey, the US Department of Agriculture found that between one-third and one-fifth of poultry farms had *negative* income.[15]

The industry is also hard on the workers who process the chickens. A May 2016 report by Oxfam America found that workers—many of them undocumented immigrants—were paid about $11 per hour to gut as many as 50 chickens per minute, and frequently developed injuries because of the repetitive motions.[16] The rates of carpal tunnel syndrome among poultry workers are about seven times the national average. Worse yet, workers are often forbidden from leaving the line to go to the bathroom, forcing many to wear adult diapers and urinate and defecate on themselves for fear of losing their jobs for an unapproved break, the Oxfam investigation found.[17] Workers and farmers alike frequently suffer infections from antibiotic-resistant bacteria from handling chickens that were fed antibiotics to stimulate more rapid weight gain. Morison, for example, said she suffered from flu-like symptoms and diarrhea about once a month. When she was examined by researchers at the Johns Hopkins Bloomberg School of Public Health, they found that she was infected with an antibiotic-resistant strain of *Campylobacter*.[18] She also developed an allergy to the antibiotic, which she believes was because of her frequent exposure to the antibiotic in chicken feed. (Perdue says it halted the practice of feeding antibiotics to its chickens to stimulate growth in 2007.)[19] Across the country, poultry workers are 32 times more likely to carry *E. coli* bacteria resistant to a commonly used antibiotic, gentamicin, than people outside of the poultry industry, according to a study that Johns Hopkins researchers published in *Environmental Health Perspectives*.[20]

Morison left the industrial poultry system to escape from all this. Now she is feeling better personally and financially, and her chickens are looking healthier and living longer. Moreover, the stream next to her farm runs clearer, too. Why? Because her farm produces a fraction of the manure it used to because it has far fewer animals. Less manure is better for the Chesapeake Bay.

"Manure is no longer really an issue here on my farm," Morison said. "We rotate our smaller number of chickens through the pasture, and the pasture is able to take up the nutrients from their waste very efficiently."

The launch of Morison's business involved some major challenges—including the fact that she started off with a heavy debt load. "When our contract was terminated, we still had that mortgage to worry about," Morison said. "So everybody had to buckle up and really put everything toward the mortgage. There weren't any extras for the family. No vacations. We didn't go shopping whenever we felt like it. But we are doing fine now."

Silbergeld, who has been studying the poultry industry for decades, said Morison's brave move to become an independent farmer should be inspiring for others. "The contract farmers have gotten the rawest end of the deal—and this has been part of the industrial farming model from the very beginning," Silbergeld said.

De Young, the spokeswoman for Perdue, said it would be wrong to suggest that the company is abusing farmers. "We work with nearly 2,000 farmers across the country, many generations of families that continue to raise chickens every year," De Young said. "If it were a bad financial situation, I don't think we'd have so many growers continuing to do it."

The irony, however, is that, in the end, Perdue reversed course and followed Carole Morison's lead. The change followed a December 2015 animal abuse scandal that was embarrassing for the company: animal rights activists in North Carolina secretly videotaped workers stomping birds to death on a Perdue contract farm.[21] In response, the company announced in June 2016 that it would overhaul its animal welfare policies.[22] Perdue said it would start *requiring* some of its contract growers to install windows in their chicken houses and allow the birds to explore outside and enjoy the sunshine.

Like Morison's chickens, Perdue may finally be seeing the light.

OOKER ESKRIDGE

Piling Rocks against the Rising Sea

I T WAS JUST AFTER SUNRISE, and James "Ooker" Eskridge, a Chesapeake Bay waterman and mayor of Tangier Island, Virginia, was in a skiff motoring across the harbor during his morning commute. The soft morning light illuminated rickety crab shacks on pillars above the water and workboats chugging out into the bay.

Above it all rose a water tower painted with a blue crab on one side and a giant cross on the other, representing the two things that keep this island town of 470 people afloat: the seafood industry and prayer.

The mayor stood at the back of the boat with his hand on the throttle. He's a tall, skinny, sunburned fisherman with a blistered lip, crows' feet at the corners of his eyes, and a battered baseball cap. He's fifty-seven years old but appears ageless, with no gray in his hair.

He tied his boat to his work shed on a platform over the water. Climbing onto the pier, he reached his hand to help his guest into his office. He then introduced his political staff: four cats that work for him, protecting his tanks of soft crabs from otters.

"That's Condi Rice," Eskridge said of the first cat. "And those are Sam Alito, John Roberts, and Ann Coulter."

The names hint at his conservative politics. And yet, when he's not tending his soft crab business, he spends most of his time wrestling with a subject that most Republican elected officials try to avoid, but which he cannot: the real-world impact of climate change. Global warming is driving up sea levels and rapidly eroding Tangier and other low-lying islands in the Chesapeake Bay and around the world.

"Our main concern is the erosion problem out here—we are losing a

lot of ground to erosion," Eskridge said.[1] "The island looks totally different from the way it did when I was growing up. Even before I came along,
there were other island communities around Tangier that have completely disappeared and are now under water. That's what we don't want
to happen to Tangier."

The vanished lands include Halfmoon Island, Delph Island, Buzzards
Island, and Wild Cat Island, among many others up and down the bay.
Over the past two centuries, more than 500 islands—large and small—
have sunk beneath the waves of the Chesapeake.[2] Some held hideaways for
pirates, while others were hunting lodges for the rich, brothels for watermen, the sites of illegal boxing matches and waterside gambling dens, and
even an unusual enterprise to breed and skin black cats to sell their fur to
China. This last scheme failed when the bay froze and the cats, wisely, ran
off across the ice.[3] Other islands were simply the homes of farmers and
fishermen, or mosquito-infested scabs of marsh grass. Fifty-seven miles
north of Tangier, Sharp's Island was once 449 acres that boasted a hotel,
four farms, a steamboat pier, and a lighthouse—but now it is marked only
by a buoy bobbing in open water. Farther south, Holland Island maintained a school with more than 100 students, as well as a baseball dia-
mond, post office, and church, before a storm in 1918 drove its last residents away and its last house collapsed in 2010.

Tangier, located just south of the Maryland/Virginia line, is about 1
mile long and 3 feet above sea level at its highest point. As mayor, Eskridge frequently travels to Richmond and Washington in a lobbying battle to try to secure millions of dollars in government funding to build a
series of bulkheads and jetties to protect his island by walling it off from
the bay. A small part of his vision has been funded, to build one jetty
shielding the west entrance to the harbor. But he's frustrated that his fellow Republicans—as well as Democrats—are not more supportive of his
efforts to save his people from becoming victims of the rising seas.

"It's not like the money's not there," Eskridge said, as country music
drifted from the radio in his crab shack and Sam Alito and John Roberts
started fighting at his feet. "The money is there, but the government is
wasting so much of it. I'm not saying you don't help foreign countries.
But we spend billions of dollars in Iraq and Afghanistan, and the bulk of
the people hate our guts there. I told the [Army Corps of Engineers], 'You
know, if you would divert some of that money here, and build us some

protection, we would appreciate it much more than those guys in the Middle East. And we promise we won't shoot at you while you are working for us.' "

Like most people on Tangier Island, Eskridge is a supporter of President Trump. Eskridge told CNN after the election, "We'll take a wall. We'd like to have a wall all the way around Tangier!"[4] The president saw the interview and called Eskridge, assuring him that he should not worry about rising sea levels. "Your island has been there for hundreds of years, and I believe your island will be there for hundreds more," Trump told the mayor.

The irony is that Eskridge is literally up to his knees in sea-level rise every time storms blast a foot of water across his island. But he—like President Trump—does not actually believe in climate change or sea-level rise. Instead, Eskridge sees himself as fighting a *sinking land* crisis caused by the naturally settling geology of the Chesapeake region. Scientists have concluded that he is partially right, as the land has been slowly subsiding since the retreat of glaciers some 12,000 years ago. But his sinking land fixation ignores the bulk of the problem, which is that warming temperatures are expanding the volume of the seas. According to the US Geological Survey, the sea level in the Chesapeake Bay is rising at a rate of about 4 millimeters per year, while the land around the southern Chesapeake is settling at a rate of 1–5 millimeters per year.[5]

"There is no evidence of sea level rise here," Eskridge insisted. "You know, maybe in some places it's occurring. But out here, you know, whenever we have a persistent East wind, we have above normal tides. You hear a lot about global warming and global climate change and all this, but things run in cycles. Nothing has really changed, there."

Nothing has changed, except that Tangier Island has lost two-thirds of its landmass over the past century and surrenders another 9 acres every year, in part because of sea-level rise driven by greenhouse gas pollution. Within a few decades, the island—with its unique culture and history—will likely vanish forever.

Even if Eskridge sees no evidence of the real cause of all this, others on the island do.

"If they don't do something, we're just going to wash away," said Danny Parks, a sixty-nine-year-old waterman whose family has lived on the island for three centuries.

"This past winter alone we had *eight* floods," agreed Maureen Gott, owner of the Bayview Inn, a charming Victorian house with rental cabins and a watch cat named Storm. "They were not all real high floods. But they were enough where you had to wear boots because the water was up all around your house."

Gott is from south New Jersey, a former respiratory therapist who bought the inn and moved to Tangier in 2011 to enjoy the peace and beauty of an island that seems frozen in time.

"We didn't really realize what we were in for when we bought the place," Gott recalled. "Then we encountered our first flood just two weeks after we closed on the purchase. There was a Nor'easter, and the bay was above my knees in my front yard. My husband was in military service in Afghanistan at the time, and I called him up and told him: 'You are really lucky I didn't turn around and go back to New Jersey!' because I was shocked at how much water there was. But then I learned that this is something that is considered normal on the island."

Whether or not the islanders recognize the source of their problems, the truth is that their lands are so low that Mayor Eskridge and his people are facing a possible catastrophe, perhaps with the next hurricane. Beyond the rising seas crisis, the island is also battling a shrinking population problem, with the number of islanders dropping by two-thirds since the 1930s. And the island community is also fragile because its economy is dependent on oysters and crabs, which are precarious sources of income because of the chronic overfishing in the bay.

But like a lot of us, Eskridge draws strength by translating reality into terms he can deal with—whether or not those terms are true. Eskridge can't stop the oceans. But he can pile rocks along the shore, and so that is what he is driven to do.

His good spirits and universal kindness spring from an old-school kind of faith based on a literal interpretation of the Bible. On the front of his crab shack, near the steel frame of his crab harvesting dredge (called a "scrape"), is a large painting of a fish with the words "We believe Jesus."

"The disciples of Jesus were all fishermen," Eskridge pointed out. "In terms of the Chesapeake Bay, a lot of the regulatory folks and scientists think it's up to them to save the bay and keep crabs in the bay. But ultimately it's up to God. Of course, God can work through people. But without the Lord, there is no crabbing. He's the one who puts the crabs in the bay, and takes care of them. And He also takes care of the watermen who catch the crabs."

People from cities may not agree with everything Eskridge believes. But his faith gives him the energy to get up every day and serve his island. Many days, he uses his power not to lobby senators or plan seawalls, but just to listen to the problems of the people with hard lives all around him.

Over the course of the day during which I followed the mayor on his normal round of duties, I noticed that Eskridge gave generous amounts of time to everyone who needed it—including to inquire into the health of a bent old waterman sitting beside a diesel fuel pump. Later, he stopped to ask another about the newborn twins of the fisherman's daughter, one of whom was born with a heart murmur. When there are domestic disputes in the middle of the night on the island, people call in Mayor Eskridge. He calmly and patiently tries to sort things out, enduring the acid and anger of such situations because he believes that his neighbors are fundamentally good. His patience and tolerance are epic—personality traits that can also be seen in his quietly determined efforts to convince politicians to fund the construction of a wall around Tangier.

"Ooker is doing the best he can to save our island," said Leon McMann, an eighty-five-year-old waterman. "But we've got no *big people*—people what knows people—no *high people*—around here. So Ooker is trying what he can."

Not everyone is a fan of the mayor. "We could do just as well without him," offered Eskridge's older brother, Ivan, who is also a waterman. "He spends all of his time with his soft crabs."

It is true that Eskridge's calendar—like the lives of most islanders—is largely consumed with crabbing in the warm months and oystering in the cold (with a little bit of eeling and puffer fish catching thrown in).

When he's not lobbying in Richmond for seawalls, he's talking to state officials about his desire to keep to a minimum the regulation of watermen and the seafood industry. He bridles at a proposal, made by some scientists and environmentalists, that a moratorium on harvesting oysters is necessary because populations have fallen to a fraction of 1 percent of historic levels.

"I've heard the argument that we've got to close the rocks [oyster beds], and maybe some of the crabbing for a few years," Eskridge said. "But the watermen are dependent on working *every year*, and for a lot of people here, survival is day to day. If they close down the harvest for four or five years, what are the watermen going to do? They'll lose their boats and their homes. Most watermen feel like I do: To save the resource, if they eliminate the watermen, it's not worth it. Watermen don't give a rip about the bay if they are going to get us off of the bay."

After scattering a handful of food for his cats, Eskridge used a hatchet to chop up a fish so he could toss chunks of it to his other workmate, a laughing gull named Summertime. Then he turned to the real task at hand this morning: bending wire to attach shiny blocks of zinc onto cages that filter the bay's water before it is pumped into his crab tanks. He explained that the zinc prevents corrosion. Then he hammered strips of wood along the edges of the cages, to hold the wire mesh in place, and brushed on a fresh coat of blue anti-fouling paint.

At about 9:00 a.m., a man in sunglasses and a baseball cap motored up to the pier and hopped off. "Hey, Ooker, can I get a dozen soft crabs?" Eskridge dug into his cooler and pulled out a mess of crabs wrapped in clear plastic, and the man paid him in cash.

Soft crabs are Eskridge's specialty. He's the only waterman on the island who focuses his business solely on "peeler crabs." When he or other watermen catch blue crabs that have reddish coloring on the tips of their swim fins (a sign that they are about to shed), Eskridge plunks them into one of two dozen tanks bubbling on a platform behind his shack. Water pours into the tanks from plastic pipes, and bare lightbulbs dangle overhead on drooping wires.

From May until October, Eskridge comes out here day and night—sometimes starting at 2:30 a.m.—to monitor the molting of his crabs. He carefully watches them, separates them, and advances the crabs from one tank to the next as their coloration changes and they shed their shells. He must shift the crabs at different phases into segregated tanks, because hard-shelled crabs will eat their soft roommates the moment they are naked and rubbery. Once the crabs have slipped their carapaces, he wraps them in plastic and pops them into his refrigerator, which prevents the natural hardening of new shells. Eskridge is good at this game. He sells his soft crabs to buyers up and down the East Coast. But he grouses that his business increasingly is being hurt by the import of Asian soft crabs at much lower prices.

"The Chinese are sending all these soft crabs over here—and it's just killing us on the market," Eskridge said. "They take foreign crab meat and label it as 'local Chesapeake.' They shouldn't be able to do that."

At about 11:00 a.m., a siren shrieked from town, piercing the silence that had blanketed the harbor. A helicopter appeared, thumping over the boats as it pounded toward the town's airstrip. "Must be a medical emergency," Eskridge said.

He leaped into his motorboat, pulled a ripcord to start the engine, and sped into town, a wake spreading behind him. He tied his skiff to the pier and climbed into a golf cart, which has a sheet of floppy plastic as its windshield. There are almost no cars on Tangier Island, only electric carts, scooters, and bikes. The "roads" (really, blacktop paths) on the island are only about 6 feet wide.

With Eskridge flooring it, we bounced down the narrow path between white clapboard houses that stand close to one another. Clusters of gravestones rise from many of the postage-stamp front lawns. That's one of the first things visitors notice about the town, after its nineteenth-century feel and the lack of any national chain stores or restaurants: Tangier has a severe grave space problem. Much of the island is marshland, broken up by wandering creeks and wooden bridges that connect the homes, post office, school, two churches, three restaurants, health clinic, general store, and beach. The few lots that don't have houses on them are packed with gravesites, many with massive cement slabs on top so the floodwaters don't float the caskets away. Dozens of graves are packed, side by side, next to the church and beside the basketball court. Kids play hide-and-seek among the stones, laughing and screaming and running. The dead are always with the living on Tangier—and this connection to history plays a powerful role in its sense of identity.

The first European known to see the island was Captain John Smith, the English explorer who, in 1608, named it and nearby islands "Russel's Isles," after his physician.[6] The island later was marked on maps as "Tangier," perhaps because an English visitor thought that its long, sandy beach looked like the one at Tangier, Morocco, in North Africa. Despite the island's beauty, 36 years after Smith's visit, Tangier became something very ugly: the first prison without walls in North America during a war of ethnic cleansing. In 1644, a Native American tribe, the Pamunkey, rose up against the invading tobacco plantations and killed about 500 English colonists. In a carefully planned sneak attack, the Indians entered homes peacefully, without any weapons, and then used their hands or whatever knives or tools they could find to kill everyone in the houses. In retaliation, Governor William Berkeley launched a war of extermination against Indians in the Virginia colony. His soldiers stormed the stronghold of the Pamunkey leader, Chief Opechancanough, captured all males over the age of eleven, and shipped all the survivors to the island later known as Tangier, where they were left to fend for themselves. It is unclear whether the prisoners died on the island or somehow escaped across the 12 miles of open water to the Eastern Shore. Their fate remains a mystery. Four decades later, however, there were no Indians left on the island when it was settled in 1686 by its first English residents, John Crockett and his eight sons and their families, which brought livestock and started farms. ("Crockett" remains a common name on the island •today, along with those of other early settler families, including the Parks, Pruitts, Thomases, and Dises.)

During the Revolutionary War, Tangier Island was a hideout of Tory sympathizers whose ships raided commercial traffic on the lower Chesapeake Bay. This may have been because the Tangiermen (whose speech still bears traces of English accents) were eager to remain on good terms with whoever ruled the waves—in this case, the British Navy. During the War of 1812, the British built a base on Tangier Island called Fort Albion. From this fortress, they launched a series of violent attacks on the young republic—including burning down the White House and the Capitol, and unleashing what Francis Scott Key later immortalized as the "rockets' red glare" at Fort McHenry in Baltimore. The nerve center for these operations, Fort Albion, is now underwater, a half mile out into Chesapeake Bay—evidence of how much water levels have risen since then. Also during the War of 1812 on Tangier Island, the British trained escaped American slaves to fight for them. The invaders had enticed many in bondage to join the British army with offers of freedom. The black "Colonial Marines" from Tangier Island sacked Washington, DC, with the British forces. After the war, the English—knowing that their volunteers would be hanged—sent the freed slaves to live as farmers in Trinidad, where today there is still a community of descendants of these freedom fighters who call themselves "the Merikens."

Mayor Eskridge's family (on his father's side, with the name Eskridge) moved to Tangier Island during the Civil War. His great-great-grandfather was an officer in the Confederate Army, but his family fled their home in Fredericksburg, Virginia, when the gunfire became too close during bloody waves of attacks and counterattacks over the city. The Eskridges took a boat to the safety of Tangier Island, even though this Virginia community had voted to align itself with the North because most residents were

Northern Methodists, who were antislavery. A son of the Eskridges married a daughter of the Pruitt family, which had been farming and fishing on Tangier since the late 1700s.

From all this history has flourished a unique culture that may soon disappear. Every April, for example, is the annual Tangier Island blessing of the fleet. The local pastor tosses a wreath into the water and reads aloud all of the names of the Tangier watermen who have died on the bay over the centuries. Another example of local tradition is the annual Tangier Island homecoming celebration every August, during which locals welcome back family and friends who have left. Everyone dances until midnight, feasting on clam and oyster fritters, soft crabs, and funnel cakes.

An outsider to such parties might have a hard time understanding what people are talking about. "We have a different way of talking here on the island, and different terms for everything," Eskridge said. "For example, the word 'yuh-hear' means three things on the island—it means *year.* It means, *your ear.* And it also means *you're here.*" He explained that his nickname, "Ooker," comes from the way he pronounced the word "rooster" when he was a kid, when he was always excited to see his family's bird. His six brothers and sisters thought it was funny and a fitting name for the cocky youngest sibling.

Eskridge's golf cart sped past the island's school, which was built on stilts to protect it from the rising waters. He explained that there are 60 students in kindergarten through twelfth grade, including six seniors in the graduating class this year, about half as many as two decades ago. "It's a great place to raise kids," Eskridge said. "You can just turn your kids loose, and you don't have to worry about them, because they can't go anywhere." For example, when they celebrate prom on Tangier Island, drinking and driving isn't really an issue. Juniors and seniors dress up in their finest and then walk in a parade down Main Ridge Road, as residents line the way and cheer them on. The couples enjoy dinner at the local seafood diner, Lorraine's, and then head to the town's recreation center/gym for a dance, followed by an all-night scavenger hunt on golf carts. "It's a big to-do," Eskridge said.

Our cart rumbled over a wooden bridge that spans a creek that wanders through marshland. A rowboat was half-submerged in the muck, with a great blue heron perched on its bow.

We pulled up at a tiny white wooden building with peeling paint and an antenna on top. This is the Tangier Town Hall—Eskridge's White

House, the source of his power and authority. "The only difference is, I don't have a Secret Service," he said, before bounding up the steps and tugging open the door.

Inside the office, Eskridge headed to the window to watch the helicopter on the landing strip. Paramedics were carrying a woman on a stretcher into the aircraft. The mayor asked two women working in the office—the town manager, Renee Tyler, and her assistant, Marilyn Pruitt—what was going on. They explained that an elderly resident had experienced chest pains and so was being carried by the Maryland State Police to a hospital on the mainland. (Tangier does not have its own hospital.) Everything seemed to be under control, so there was no need for the mayor to personally intervene, as he often does—to perform CPR, for example, or personally carry the woman to safety.

With the crisis under control, the town manager, Tyler, glanced down at the tall stack of bills on her desk. The top one was ripped open.

"It's a tough job," said Tyler, who reports to Eskridge. "It's a lot of responsibility, trying to keep the town in order."

She explained that the town has an annual budget of about a half million dollars per year. She and her assistant are two of the five municipal employees, with the others being a wastewater manager, a police officer, and a maintenance man. She headed into the next room and yanked open a cardboard box full of colorful T-shirts. The shirts proclaim "Save Tangier Island" on one side and "#TangierIslandLivesMatter" on the other, followed by "I Refuse to Be a Climate Change Refugee."

Tyler explained that the town is selling the shirts for $20 each to tourists who visit, arriving mostly by boat on weekends in the summer. So far the effort had raised only about $1,400 toward saving the island. That's a far cry from the tens or hundreds of millions of dollars that might be needed to build seawalls all the way around the island to temporarily protect it from the rising waters.

"Yes, we have a ways to go—but we are taking a stand and fighting," Tyler said. "We are not going down easy. A lot of people think, 'Why don't those people on Tangier Island just move somewhere safer? It would be cheaper to solve the problem that way.' Well, we don't want to move! Let the people who say that leave their houses and move. Tangier Island is a gem, rich in history and cultural heritage. This is where we live and this is where we want to die."

After the mayor discussed some business matters with his staff, he headed back outside to his golf cart and drove home for lunch. On the way, he explained that his wife, Irene, is also from the island and the child of a waterman. They met when they were fifteen, dated in high school, and married immediately after graduation in 1977. They had two sons and then adopted four girls from an orphanage in India. "My wife had always wanted a girl. So after she had some complications giving birth to our second son, we looked into foreign adoption," Eskridge said. "Kids are the health of your island. If you've got a community with no young people in it, your community is dying. So when I talk to young couples around here, I tell them: 'You've got to marry early and have lots of kids.'"

After lunch, Eskridge drove back to the town docks for an important daily ritual on the island: the 1:15 p.m. arrival of the mail boat from Cris-field, Maryland, about an hour away. About 30 townspeople gather here every afternoon to wait. Traditionally, the mail boat was the island's only connection to the rest of the world—although, of course, now people on Tangier have phones and the Internet, too. Folks gossip, hug, and joke as they wait for their mail boat to arrive. When it pulls up, a man in a T-shirt unloads a parade of packages, including bundles of paper towels, boxes of soap, groceries, and cat food.

Watching the packages is John Charnock, the island's sole police offi-cer. His job is not exactly like working the police beat in Baltimore. There hasn't been a murder here since the early 1970s, when two watermen got into a scrape in a crab shanty. But Charnock said he keeps his eyes open. "There is a little crime here—a prescription drug problem on the island," he noted. "People will get their prescriptions filled for pain killers, and then sell half of them. There is also a little bit of heroin and cocaine that trickles in, too. It is very tough working as a police officer here, because everybody on the island is either my family or a close friend."

As he talked, a dark line of storm clouds rolled in from the west, shad-ing the harbor so that the crab shacks glowed white before the gathering darkness. Eskridge climbed into his skiff and motored back out to his shack. Moments after he arrived, rain started hammering the roof of his building. In the downpour, he saw his oldest son, James—the only of his six kids who followed his father into the fishing business—roar past in his workboat, with bushels of crabs piled high near the stern.

Eskridge reflected that it can be a dangerous profession, as he watched his son disappear into the storm and the rain fell harder.

"It's ironic—the Chesapeake Bay, which supplies the livelihood of the island, at the same time can be pretty harsh and can take your life if you are not careful," Eskridge said as the harbor boiled and the wind rose. "The same bay that has provided life for the island for hundreds of years is now the problem. The Chesapeake Bay is trying to gobble Tangier Island up."

The Wildlife

NEXT, WE SPEND TIME with the nonhuman life in the Chesapeake. The bay's thousands of miles of sheltered inlets, as well as its blending of salt water with freshwater, make it a breeding ground and hideaway for a spectacular array of creatures.

The oyster built the foundation of the bay with its body. But its tale is a tragic one, with overharvesting ripping the Chesapeake's base out from beneath it. Recently, however, Maryland has created sanctuaries to protect oysters, and they seem to be working—although watermen are determined to dredge even in these no-fishing zones.

Blue crabs are iconic to bay culture, but there is an ugly truth to the beautiful swimmers. These cantankerous cannibals don't give a pinch about pollution—until they lose the underwater grasses that shelter them from other crabs and fish.

Striped bass are the rock stars of the bay restoration effort, with a moratorium on fishing for "rockfish" sparking a spectacular comeback. But, since then, disturbing questions have surfaced over why so many have a mysterious wasting disease.

American eels are backward fish that breed near the Bermuda Triangle and then—unlike most ocean-born animals—spend their lives far inland.

Atlantic sturgeon were long believed to be extinct in the bay. But then a secret colony was discovered clinging to life in the James River in Virginia. A biologist has been trying to play matchmaker to these dinosaur fish, but his amorous efforts hit the rocks.

OYSTERS

Pearl of an Idea: Ban the Oyster Dredge

T HE SUN BLAZED DOWN through feathery piles of clouds onto a river, illuminating a hill of oysters on a barge. A man with a hose blasted shellfish off the deck into the waters of Harris Creek, a tributary to the Choptank River on Maryland's Eastern Shore. Almost 30 million juvenile oysters were being planted atop an artificial reef of granite and old shells. The planting was part of one of the largest and best-protected oyster restoration efforts ever attempted in the United States. Harris Creek is ground zero for Maryland's controversial new strategy for boosting the bay's severely depleted oyster population. The key is that all of the oysters are being planted inside a 4,500-acre sanctuary, into which watermen are never allowed with their harvesting equipment.

"We are trying to bring back the oysters for the ecosystem services that they provide for the Chesapeake Bay," said Stephanie Westby, a biologist with the National Oceanic and Atmospheric Administration.[1] "And by ecosystem services, I mean the tremendous habitat they offer for fish and crabs, and also their incredible filtration capacity."

Eastern oysters (*Crassostrea virginica*) are remarkable: tiny wastewater treatment plants, each of which can filter 50 gallons of water per day. They are also the bay's greatest paradox: ugly to look at and yet delicious to consume; rocklike and yet soft inside; scum-suckers and pearl makers. Oysters are also bold sexual voyagers—starting out bisexual, becoming male during their first year of life, and then switching to female. Oyster reefs are the sites of spectacular orgies in the spring, when warm temperatures stimulate milky clouds of sperm and eggs that join to make millions of larvae. Oysters use their own graveyards as brothels. The larvae of oysters must attach to the shells of dead oysters—or to living oysters, or other hard surfaces—to keep themselves up out of the mud, or they will suffocate. Fish and crabs also use oyster reefs as shelter. For this reason, removing oyster reefs is the Chesapeake Bay's unholy trinity of troubles: (1) taking out oysters eliminates water filters that clean the estuary, (2) harvesting oysters scrapes away the foundation for future generations of oysters, and (3) dredging for oysters flattens the homes of many other bay-dwelling species.

Oysters were once so plentiful in the Chesapeake that the ships of explorers ran aground on their reefs.[2] Both colonists and Native Americans feasted on the mollusks, with the density of their shells heaped in Indian trash piles a sign that they have been a favored food since the dawn of human settlement. But oystering was not really an industry until after a Frenchman named Philippe de Girard invented the tin can in 1810. Before canning, oysters could not be stored for long or transported far to market, because they would spoil, especially in the summer. But inside cans, oysters became portable protein packets—the first fast food. Oyster harvesting exploded with the growth of canneries in the decades after the Civil War. The rapidly expanding railroads allowed carloads of canned Chesapeake oysters to be shipped to everyone from cowboys in the West to captains of industry in Manhattan.

The advent of oyster dredging—the dragging of heavy metal rakes with bags across the bottom—sparked a gold rush in the nineteenth century. Clashes erupted between dredgers and watermen who harvested more slowly with pincher-like devices called tongs. Maryland banned oyster dredging in 1820, because the practice ripped up reefs, but then later loosened the rules. In 1865, Maryland passed a law that required watermen to first obtain permits every year before they could dredge for oysters. The state's attempt to enforce this rule sparked a violent backlash from watermen, igniting what were called the "Oyster Wars" of the late nineteenth century.[3] Armed state police in the newly formed Maryland Oyster Navy (the forefather of today's Maryland Natural Resource Police) frequently traded gunfire with heavily armed watermen who were harvesting illegally. During the peak of this Wild West on the bay, in the 1880s, about 50,000 watermen worked in the industry, hauling up about 17 million bushels of oysters a year. (It was no coincidence that this was also roughly the same period of time in American history when, in the real "Wild West," bison were slaughtered nearly to extinction and carrier pigeons were massacred in the Midwest.) The amount of meat pulled from the Chesapeake was so massive that journalist H. L. Mencken renamed the bay an "immense protein factory."

Permanent ecological damage was caused by those cashing in on the bay's once-expansive reefs. In 1891, Maryland's oyster commissioner, Dr. William K. Brooks, raised alarms about overharvesting. "Everywhere, in France, in Germany, in England, in Canada, and in all northern coast states [of the United States] history tells the same story. In all waters where oysters are found at all they are usually found in abundance, and in all of these places the residents supposed that their natural beds were inexhaustible until they suddenly found that they were exhausted," Dr. Brooks wrote.[4] He was right. But, tragically, his arguments for the creation of oyster sanctuaries and a shift from wild harvest to oyster farming were ignored for more than a century (and are still ignored by many today). For decades, elected officials repeatedly shut their ears to biologists and listened instead to the demands of a small but politically influential watermen's lobby. The watermen fought regulation—first with guns, and later with a white hat image with the public, as they were often portrayed in the press as the representatives of the "true" Chesapeake men. By 1930, the bay's oyster harvest had plummeted by 75 percent from its high in 1880; by the 1990s, it was down more than 99 percent. In addition to the damage caused by overharvesting, oysters were also smothered by pollution (especially sediment from farms and construction sites) and, after the 1980s, periodically ravaged by parasitic diseases called MSX and Dermo.

In an effort to rebuild the bay's most important species, programs funded by the federal and state governments have planted more than five billion oysters in the bay since the 1990s. But most of the early oyster restoration projects failed. This was in part because authorities planted the bivalves in areas that were open to harvesting, and so watermen quickly scooped them out and sold them. This taxpayer-funded "put and take" fishery did nothing to restore the Chesapeake Bay's health and was a wasteful cycle that was promoted by elected officials and nongovernment organizations (such as the Oyster Recovery Partnership) that wanted to appear simultaneously pro–seafood industry and pro-bay. The shell game did not work. Although an agreement among Chesapeake Bay region states in 2000 set a goal of increasing the oyster population 10-fold by 2010, the population actually dropped by about 70 percent over this decade, according to an estimate by University of Maryland Center for Environmental Science biologist (UMCES) Michael Wilberg.[5]

During these lost "put and take" years, the numbers of oysters in the bay fell to a near-extinction level of *one-third of 1 percent* of historic levels.[6] Stop and think about that number for a moment. Imagine a forest of 1,000 trees. Then cut down nearly all the trees, leaving only three in a denuded landscape. You couldn't call it a forest anymore—just as you can't really say that oyster reefs remain in Chesapeake Bay. Trees are like oysters, in that they both clean the environment around them. Trees absorb carbon dioxide and release oxygen; oysters suck in nitrogen and release filtered water. And just as important, both oyster reefs and forests provide the shelter and nesting grounds for hundreds of other species. Without them, you have a wasteland.

The fact that oysters, in their severely depleted state, are still being harvested from the bay today is shocking and outrageous, if you look at the big picture. The government would surely halt the hunting of any animal species—elk, deer, bird—if more than 99 percent of its members were killed. In most other places around the world where oyster reefs once existed and then were scraped down to nothing, fishermen shifted from wild harvest to oyster farms, which is where most oysters sold in restaurants today come from. The same transition has started in the Chesapeake, but it must be accelerated and enhanced with a moratorium on wild oyster harvesting, before all the wild oysters are gone.

This is what Dr. William K. Brooks recommended way back in 1891, and it still makes sense today. But it has not happened. So what has Maryland done in the face of the great Chesapeake oyster crash?

In 2003, Maryland governor Robert Ehrlich proposed introducing an exotic species of hardy, fast-growing oysters into the bay. The idea was that the Asian oyster, *Crassostrea ariakensis*, would be resistant to the Dermo and MSX diseases (more about them in the next chapter) that since the 1980s had been contributing to the decline of the native Chesapeake oyster.

In 2009, Governor Martin O'Malley's administration rejected this idea after scientists concluded that introducing yet another invasive species into the bay was too risky and could allow the exotic oysters to outcompete— and kill off—the remaining native oysters. Instead, the O'Malley administration charted a new direction for oyster policy. Under O'Malley, the state protected 24 percent of the remaining oyster reefs in the state's portion of the bay, imposing sanctuaries over 8,563 acres.[7] That was more than two and a half times the 3,500 acres of scattered, barren bottom that had been shielded previously with no-harvest zones. Of course, protecting *all* of the remaining oyster reefs—instead of just one-quarter of them— would have been even better for the oysters and the health of the Chesapeake Bay. But policy makers feared a backlash from the watermen's lobby and Eastern Shore lawmakers. The watermen's lobby pushed back hard, anyway, even though the O'Malley administration tried to soften the blow and help oystermen transition to a new way of doing business, by growing oysters in underwater cages as part of aquaculture enterprises. The O'Malley administration offered low-interest loans to watermen to help them buy the necessary equipment for aquaculture. Lawmakers also approved the leasing of bay bottom in Maryland for private oyster farms. In addition, the state ramped up the growing of oysters in hatcheries in a lab at the University of Maryland Center for Environmental Science at Horn Point, with the baby oysters planted mostly inside the new sanctuaries. And, of course, the state continued to allow watermen to harvest wild oysters from three-quarters of the bay.

The initial focal point for the new approach to rebuilding oyster populations was Harris Creek. "What we are doing now in Harris Creek is totally different," said Mike Naylor, shellfish program director at the Maryland Department of Natural Resources, in May 2013. Instead of a "put and take" fishery, he said, "This is put and leave alone, and allow to expand."

The new strategy was focused on helping the oysters—not the oystermen, who received most of the attention in prior efforts. However, the idea was that watermen would benefit more in the long run through this method, because protected reefs tend to generate more oysters, fish, and crabs that spread out to populate a wider area. Watermen would benefit from this spillover, because the state would continue to allow harvest in most of the bay.

The experiment worked. Between 2010 and 2014, the estimated mass of oysters in the bay more than doubled, before falling off only slightly in 2015, according to the Maryland Department of Natural Resources.[8] In the 2013–14 season (which runs from October until March), more than 1,100 watermen harvested 433,000 bushels of oysters worth $15 million in Maryland. The totals for 2014–15 declined only slightly, to 389,000 bushels. Those numbers remained a tiny fraction of the totals from the early 1980s, let alone the nineteenth century. But still, the 2014–15 oyster harvests represented an almost fourfold increase over the 101,000 bushels harvested in 2009–10.[9] Two strong years of oyster reproduction in 2010 and 2012, likely encouraged by good weather conditions (low rainfall and therefore high salinity levels in the bay), played a role in that surge.[10] Oysters produced by aquaculture businesses on the bay also rose impressively over this time period, with more than 100 oyster farming enterprises now in Maryland, compared to only three in 2010.

"I think it's the beginning of the next revolution for the seafood industry," said Johnny Shockley, a former oyster harvester who opened up an oyster farming business, the Hooper's Island Oyster Aquaculture Company, with his father and son in 2010.[11] "Our production is increasing by a rate of 500 percent per year right now. And the markets are just continuing to build. We are reinventing the oyster industry based on sustainable production."

Part of the reason for the improved reproduction of wild oysters in the Chesapeake appears to be Maryland's expanded sanctuaries. A study by the Maryland Department of Natural Resources in July 2016 confirmed that oysters inside the state's new oyster sanctuaries were multiplying (although sometimes targeted by poaching), while the shellfish outside the protected areas continued to decline.[12] Oysters also seem to be evolving some resistance to the parasites that cause the diseases MSX and Dermo, allowing more oysters to survive longer.

The increase of the past few years is encouraging—but should be kept in perspective. Oyster numbers remain near 1 percent of historic levels and far below the targets set by the bay states in the 2000 Chesapeake cleanup agreement for the year 2010. Moreover, a change in rainfall and an accompanying shift in the concentration of salt water in the bay could easily wipe out the recent gains, in part by creating conditions that encourage the diseases MSX and Dermo. (Dry conditions and more salt water in the bay increase the activity of the MSX and Dermo parasites. This is explained in more detail in the next chapter.) We also need to recognize that recent growth in oyster population is happening in a very small percentage of the habitat where oysters once lived (mostly the saltier southern bay). There is no longer any significant oyster harvest north of the Bay Bridge, because there are almost no oysters left there.

In 2015, the fragile progress of the bay's oysters was threatened by the Maryland Watermen's Association. The watermen lobbied Governor Larry Hogan's administration to open up the new sanctuaries to power dredging and urged the state to stop building oyster reefs in protected areas. This lobbying effort succeeded in convincing the administration to temporarily halt the construction of a new oyster restoration project, using recycled shells, in the Tred Avon River in 2016.[13] Why? Because watermen were angry that the state was planting shells in a no-harvest zone, from which they could not profit.

"These new oyster sanctuaries . . . I don't see how they are benefitting anyone," complained Robert T. Brown, president of the Maryland Watermen's Association. "Hogan is a businessman. Maryland is a big business, and the seafood business is a big business, and you've got to run it like a business."

In January 2016, the Hogan administration bowed to this pressure and opened up 10 "harvest reserve" areas.[14] These semiprotected zones had been designed for oyster planting and then harvesting on a rotating basis, every few years. In February 2017, the Hogan administration proposed eliminating 976 acres (or about 11%) of O'Malley's oyster sanctuaries.[15] More broadly, watermen pushed the state to expand the areas across the bay open to power dredging. Power dredging, as mentioned previously, is the highly destructive method of using motorboats to scrape oysters from the bay's bottom with a metal rake-like device and bag.

Although banned for more than a century because of its destructive power, power dredging began to be legalized by Maryland in certain small parts of the bay in the early 1980s, and then the state opened more areas to dredging in 1999, 2003, and 2013. From 2010 to 2015, as oyster populations rose, the number of watermen power dredging skyrocketed from 418 in 2010 to 723 in 2015, according to the Maryland Department of Natural Resources.[16] About half the oysters caught in Maryland are pulled up with power dredges, with the rest caught with metal scoop-like devices called tongs, or by divers.

Watermen—without scientific evidence—argue that power dredging increases oyster populations and improves the health of the bay. "It's almost like if you have a garden," said Jim Mullin, executive director of the Maryland Oystermen Association. "You've got to turn the soil over, right? So when you have an oyster bar, you've got to go in there every year and you've got to bring those shells up to the surface so the oyster spat [baby oysters] can catch on it. The purpose of power dredging is to get those shells out of that mud and sand."

It's a claim that lacks data—or even common sense—to back it up. Power dredging doesn't just "get those shells out of the mud" (which might be helpful, if they were being lifted a few inches off the bottom). Dredging rips the oysters clear out of the bay, so they end up on a plate and then in the trash. In that way, the claim that dredging helps oysters is like the assertion by loggers in the 1980s that clear-cutting forests helped wilderness areas by preventing acid rain. Back then, loggers claimed that the acidity in streams was leaking from the needles of pine trees, instead of from the true source: sulfur dioxide pollution from coal-fired power plants.[17] So cutting down trees would help the environment, the loggers argued (incorrectly). With the same logic, oystermen assert that pulling oysters out of the bay helps the bay.

"Power dredging does two things: it flattens the reefs, when we need the opposite, we need three dimensional reefs," said David Goshorn, assistant secretary for aquatic resources at the Maryland Department of Natural Resources. "And dredging also re-suspends sediment, and the nutrients associated with that can be bad for the bay. So from an oyster restoration perspective, power dredging is not the answer."

So what is the answer? In my view and the opinions of others, the answer is more oyster restoration projects in protected areas like Harris Creek. But we also need to go further and expand these kinds of efforts all over the bay. In other words, the whole Chesapeake should become a sanctuary—a 200-mile-long protected zone, in which more oysters are planted but harvest is illegal until the species can recover. Power dredging should be banned permanently because it is just too damaging, like driving bulldozers through state forests.

Michael Wilberg, biologist at the University of Maryland Center for Environmental Science, advocated for a moratorium on oystering in a 2011 scientific journal article.[18] "The magnitude of the decline raises concerns about potential for continued loss of natural oyster beds throughout much of Maryland waters," Wilberg wrote. "Therefore, we recommend a moratorium on fishing until reefs and self-sustaining populations are restored."

There are two main arguments against a moratorium. The first is that the income of watermen would suffer. But this is short-term thinking, because their income from the harvest of oysters will eventually be zero if their overfishing does not stop. More importantly, oysters serve a far more important role in the ecosystem than simply being a source of cash for people, as they also clean the bay and provide homes for many other forms of life. The second argument against a moratorium is that it's already too late: there are too few shells left on the bay bottom to serve as a foundation for the future reproduction of oysters. The logical response to this claim is that Maryland and Virginia should expand their use of rock or cement as substrate for oyster plantings and continue to plant oysters on top of recycled shells obtained from restaurants or seafood processing plants. Then the oysters should be left alone, so that any future reproduction can serve as the base for more rebuilding. Now that we realize we're in a hole, we must stop digging.

Some argue for a more gradual approach—that we should, as a politically pragmatic first step, continue our experiment with preserving a quarter of the bay's reefs—and then, in a decade or two, if the data demonstrate that these sanctuaries work, slowly expand them from 24 percent to, say, 30 percent of the remaining reefs. The problem with this "go-slow" approach is that the theft of oysters from sanctuaries today is so rampant (a problem that regulators have documented,[19] including poaching from the Harris Creek sanctuary) that we'll never really know how well sanctuaries work until we ban the dredges that are used for poaching. Enforcement of sanctuaries will only be possible, given the state's shrinking number of natural resource police officers, when we get all oyster dredging equipment off the bay. We cannot save the bay without saving the oyster.

DERMO AND MSX

The Parasite Paradox

O NE OF THE MOST mysterious creatures in the Chesapeake Bay is a microscopic, single-celled animal—a protozoan—that swims about, propelled by a pair of whip-like flagella. Under a microscope, *Perkinsus marinus* looks harmless, like a tiny bubble. But when oysters suck in water and accidentally ingest these bubbles, they burrow into the oyster's fleshy organs and seize control of them for their own selfish purposes. The parasite uses the oyster's body to multiply its own offspring, leaving the oyster pale and shriveled. When the host defecates or dies, the bubbles swarm to neighboring shellfish. This disease, called Dermo, has been slowly killing between 10 and 20 percent of the bay's oysters in recent years.[1] But in the 1980s and 1990s, it was much worse—as Dermo killed more than half of the bay's oysters in some years.

Some parasites in the Chesapeake Bay are clearly invasive species— aliens from Asia, the American Midwest, or elsewhere which were transported here inside the ballast tanks of ships or by other means. In this category is another oyster parasite, *Haplosporidium nelsoni* (source of the oyster disease called MSX, discussed below). But scientists call *Perkinsus* "cryptogenic," meaning that its origin remains unknown. Some argue that *Perkinsus* has always been lurking in the Chesapeake but only caused widespread death in the 1980s when drought conditions, high salinity, pollution, and low-oxygen conditions combined to make it more virulent. Other researchers suspect that the parasite is from Louisiana and the Gulf of Mexico.

The theory that *Perkinsus* is a Cajun invasion goes back to just after World War II. In 1946, Louisiana watermen were horrified that their beloved Eastern oysters (*Crassostrea virginica*, the same species native to the

Chesapeake Bay) were dying off. They filed lawsuits against the oil industry, which was increasingly drilling atop their oyster harvesting beds on
the edge of the Gulf of Mexico. Scientists investigated the withering shellfish and discovered the disease. At first, researchers misidentified the cul-
prit as a fungus, and then as a slime mold—before they finally figured out
that it was actually a single-celled parasite. The researchers concluded that
Dermo, and not the oil rigs, was the real cause of the die-offs. Scientists
then found evidence of the disease in Louisiana oysters dating back to the
1920s which had been preserved. In 1951, other biologists also discovered the parasite in Florida, South Carolina, and Virginia's James and
Rappahannock Rivers in the Chesapeake Bay watershed. This suggested
the possibility that the parasite was being spread, perhaps in oysters being
transplanted from the gulf up the East Coast for oyster farming businesses. A Dermo disease outbreak in Delaware Bay in 1955 was pinned on
infected oysters moved from the Chesapeake. And in the 1980s and 1990s,
the disease spread to New York, New Jersey, Massachusetts, and Maine.[2]

Other researchers have offered an alternative theory. What if *Perkinsus* didn't invade—but was always in the Chesapeake Bay and other waterways, and it was just the diagnosis of the disease that spread? That could
be true, but if so, it raises another question: if the parasite has been in the
bay for millennia, why did it only start wreaking serious havoc on the
Chesapeake's oysters in the 1980s?

This remains a mystery. But Denise Breitburg, senior scientist at the
Smithsonian Environmental Research Center, is shining a light on this
enigma. She investigated the disease and concluded that, in some cases,
Perkinsus is more deadly in low-oxygen zones, typically caused by pollution.[3] Nitrogen and phosphorus from fertilizer, sewage treatment plants,
and other sources feed algal blooms, which cause wild swings in oxygen
levels in shallow parts of the bay between day and night. These nighttime
low-oxygen "dead zones"—which scientists call hypoxia—weaken the oysters' immune systems. And weakened disease-fighting capacity seems to
have made the oysters more susceptible to infection with Dermo from
parasites that may have been haunting the bay for a very long time. "We
definitely do have a link between hypoxia and the amount of oyster disease," Breitburg said.[4]

Scientists have known for many years about low-oxygen zones in the
deep parts of the bay. Nitrogen and phosphorous pollution spur the growth
of algae, whose decay after death sucks oxygen out of the water, especially

in deep areas. But oysters don't grow naturally in the deep parts of the bay. What Breitburg documented was a new kind of "dead zone" that appears only at night in shallow inlets, where oysters do grow.

"Basically, during the daytime, phytoplankton—plants in the water —are producing oxygen through photosynthesis," Breitburg explained. "Throughout the day, everything in the water is respiring. They are breathing like you and I would, and so they are consuming oxygen and also producing carbon dioxide. But at night, when oxygen is not being produced, respiration [the consumption of the oxygen] is what's dominating. And so oxygen concentrations go down at night."

These low levels of oxygen after sunset—made worse by more pollution—are associated with higher Dermo infection rates in oysters. Another factor in Dermo is the saltiness of the water. The parasite lives and thrives in salt water, not freshwater, and so it spreads in years when there is little rain, and therefore less freshwater in the bay and more salt water from the ocean, according to Mike Naylor, former director of the shellfish program at the Maryland Department of Natural Resources. Naylor said that it is true that many fewer oysters have died from Dermo over the past decade, compared to the more than 50 percent killed in previous decades. But that lower death rate might not be because of a drop in pollution. It might simply be a reflection of more favorable rain conditions in recent years. "It is important for everyone to remember that Dermo disease becomes very lethal when we have droughts—particularly prolonged droughts, and we have not had a significant drought in the Chesapeake Bay area in 11 years," Naylor said in 2015.[5] "That is exactly the amount of time since the last very significant Dermo mortality event."

Several puzzles remain, however. In centuries past, did droughts also cause spikes in *Perkinsus* abundance and die-offs among oysters in the bay? Maybe not, because the lower amounts of pollution and smaller "dead zones" back then meant that there was less stress on the oysters' immune systems. Some researchers also speculate that *Perkinsus* may have mutated and suddenly became more virulent in the 1980s and 1990s. And in response to these especially deadly years, oysters might have evolved resistance to the parasite that we are seeing today. If so, this might help explain why the estimated oyster population in the bay—while still extremely low—has doubled over the past five years.[6] A new generation of oysters may be adapting to tolerate Dermo (and also flourishing in favorable rain conditions).

Oysters also appear to be evolving resistance to Dermo's partner in crime—the other major parasite, *Haplosporidium nelsoni*, which causes the disease MSX. MSX is killing far fewer oysters today than it did in the 1950s through 1990s. In Virginia's York River, for example, fewer than 5 percent of oysters monitored in 2010 were dying from MSX, compared to more than 50 percent a decade earlier.[7]

Unlike *Perkinsus, Haplosporidium nelsoni* is clearly an invasive species—an outsider from Asia. Under a microscope, this parasite looks like an egg wearing a hat. Despite its harmless and oddly dapper appearance, however, *Haplosporidium nelsoni* is aggressive—far more fast-moving than *Perkinsus*. The invader attacks an oyster's gills and kills the bivalve quickly. When scientists first detected this parasite sweeping through oysters in the lower Chesapeake Bay in 1959, they named the disease it causes MSX, which stands for "multinucleate sphere X." Detective work by biologists deduced that the parasite was most likely brought to the bay by oyster farmers trying to transplant a Pacific variety of oyster, *Crassostrea gigas*, from Japan or the Pacific Northwest into the Chesapeake during the 1950s.[8] The new oysters likely were carrying the microscopic parasites but had developed resistance to them over many years. The Pacific oysters perished in the Chesapeake, but their passengers were spectacularly successful, especially in the saltier, southern part of the bay. Following the waves of mass deaths, the oysters that survived tended to be those that had some resistance. And so over time, the bivalves appear to be gradually evolving an ability to withstand the parasites, just as the Pacific oysters did.

Looking to the future, it may be that the Eastern oysters will be able to overcome this invasion and survive in their native habitat of the Chesapeake Bay—if only humans, the ultimate predators, would leave them alone.

BLUE CRABS

The Ugly Truth about the Beautiful Swimmers

E VERY JUNE, my family makes a pilgrimage to a rickety pier on Virginia's Eastern Shore. My daughters toss chicken necks on strings into the foot-deep water and watch as tiny fish swarm, tugging and ripping at the meat. After a minute or two of watching the mummichogs, Jane sees her real target approaching across the muddy bottom: a blue crab, massive as Goliath in comparison to the fish. It strides up to the bait with its claws raised. "Come on. *Come on!*" she whispers, lying on her stomach on the broken wooden planks with the string in her fingers. If the crab is thinking about the risks versus the rewards in this situation, it doesn't think very long. It latches onto the bait. Jane pulls slowly and gently upward on the string, and the crab's greed is so intense that it does not let go, even as she lifts it clear out of the water. Goliath is hooked only by desire. That night we steam Goliath and a dozen of its friends in National Bohemian beer and Old Bay Seasoning and enjoy a feast.

Blue crabs are an object of cultural devotion in the Chesapeake region. They are odd objects for adoration, however, because they seem to contain more malice than meat. Blue crabs are pugnacious, violent, edgy characters that will eat just about anything, from worms and clams to rotting fish and even their own children. Yet, we love them! So tasteful. For a long time, it was assumed that blue crabs were relative newcomers to their position at the apex of Chesapeake cuisine, only climbing to their place of prominence because of their status as the last species standing, after oysters, shad, and terrapin were eaten nearly into nonexistence. This assumption was based, in part, on archaeological research that concluded that Native Americans were big consumers of oysters, but not crabs, as judged by the trash heaps they left around their campsites. These heaps—

Workboat returning from the Chesapeake Bay. Three voluntary, state-led bay cleanup agreements in 1983, 1987, and 2000 failed to produce any improvement in the bay's overall health between 1986 and 2010 because they did not include regulations that would have reduced pollution or protected wildlife. In 2010, the Environmental Protection Agency imposed a new federally led system of pollution limits that finally appears to be working—but is now threatened by President Donald J. Trump's dismantling of the EPA.

Maureen Gott, owner of the 1904-built Bay View Inn on Tangier Island, Virginia, a historic fishing town rapidly being consumed by rising sea levels. "We didn't really realize what we were in for when we bought the place," said Gott, who is from New Jersey, holding her cat named Storm. "This past winter alone we had eight floods."

Ooker Eskridge, a waterman and Republican mayor of Tangier Island, has
been lobbying for government funds
to build seawalls to protect his town
from the rising seas. But he doesn't
believe in sea-level rise or the climate
change causing it. Nor does he put
much faith in science or government.
"A lot of the regulatory folks and
scientists think it's up to them to save
the bay. But ultimately it's up to God,"
Eskridge said.

Rare, untouched stretch of Chesapeake Bay beach, as it might have
looked when English explorer Captain
John Smith set foot there in 1607.

Camping on an island near Old Plantation Creek on Virginia's Eastern Shore during a kayak expedition led by veteran environmental educator Don Baugh and Chesapeake Bay author Tom Horton.

Tom Horton, naturalist, kayaker, and author of *Bay Country, Turning the Tide, Chesapeake: Bay of Light,* and several other books about the Chesapeake. "One of the things that makes the bay special is that it's a place where land and water intertwine so extensively and intimately," Horton said. "These margins of land and water—the mudflats, the seagrass beds, the oyster rocks, the tidal wetlands—are among the most productive habitats on earth."

(*top*) Carole Morison raised chickens for Perdue for 23 years on Maryland's Eastern Shore. Then one day she decided she'd had enough of the company's rules, and she opened her own business selling the eggs of free-range chickens to Whole Foods. "I'm a completely different person now," Morison said. "I really enjoy the independence of being a business owner. It's more American."

(*bottom*) Fred Tutman, the Patuxent Riverkeeper, in front of the Chalk Point coal-fired power plant, one of about 10 polluting businesses, including Walmart, that he has sued to protect his home waters. "These are David and Goliath fights," Tutman said. "But the truth is, I believe in the power of one. I believe in the power of dissent. I believe you, as an engaged citizen, have a lot of avenues to take in hand the problems of your local community and your local waterway."

Chuck Fry, a dairy farmer, is president of the Maryland Farm Bureau. He fights for fewer environmental rules for farms—even though agriculture is the single largest source of pollution in the Chesapeake Bay. "What really needs to happen is that the environmental community needs to really listen to the farmer," Fry said. "Or we are just going to have to go down the path of farmers need to do what they need to do, and regulators need to do what they need to do, and somewhere in the middle there is going to be a clash."

(*top*) Veteran waterman Dorsey Shockley (*right*) joined his son, Johnny (*left*), and grandson, Jordan, in transforming the family into oyster revolutionaries. Instead of just dredging oysters from the bay bottom, the Shockleys opened an oyster farming and technology business. Their Hooper's Island Oyster Aquaculture Company sells aquaculture systems over the Internet and 20,000 cage-grown oysters per week to about 100 high-end restaurants from New York to Richmond.

(*bottom*) Robert T. Brown Sr., president of the Maryland Watermen's Association, lobbied Governor Larry Hogan's administration for the elimination or reduction of oyster sanctuaries, which protect the increasingly rare shellfish from harvest. "Maryland is a big business, and the seafood business is a big business, and you've got to run it like a business," Brown said. About 1 percent of the bay's historic oyster populations remain.

Steamed blue crabs on the docks at
Fishing Creek, on the Eastern Shore.
Maryland and Virginia boosted the
population of "beautiful swimmers"
in 2008 by imposing restrictions on
catching of females. But since then,
numbers of crabs have once again
become erratic, in part because the
complex lifecycle of the species is
dependent on the whims of weather.

Baltimore's harbor. Despite its great
beauty, culture, and history, "Charm
City"—like many older American
cities—is built atop infrastructure that
was badly neglected during much of
the twentieth century. The city's leaky
and overwhelmed sewer system, for
example, routinely dumps raw sewage
into the Inner Harbor during rains.

The Blackwater National Wildlife Refuge is a 28,000-acre nature preserve on Maryland's Eastern Shore that is sometimes called the "Everglades of the North." It is threatened from one side by encroaching suburban sprawl and from the other by rising sea levels that swallow 300 acres of marshland every year.

The James River is awash in history.
The waterway was the port of entry
for both slavery and democracy into
America; the birthplace of corn,
cotton, and iron mills that marked the
beginnings of the nation's industrial
revolution; and the artery of commerce that fed the Confederacy and its
heart, Richmond.

At the edge of water and land in Dorchester County, Maryland. An island floats in mist at sunrise.

called "middens"—seldom contained any blue crab shells. In the past few years, however, scientists at the Smithsonian Environmental Research Center have discovered that these historic sites actually hold evidence of blue crab feasts—tiny particles of crab shells, which had previously been overlooked.[1] As it turns out, crab shells dissolve much more quickly than oyster shells, in part because acid rain from coal-fired power plant pollution breaks down crab shells more readily. But the small fragments that remain—specifically, the tips of crab claws—suggest that the local tribes gobbled up considerable quantities of blue crabs.[2] More importantly, the crabs back in precolonial times were as much as *twice* as large as the average blue crabs today. The researchers determined this by measuring the size of the claw tips and comparing them to the tips of claws today, according to Anson "Tuck" Hines, director of the Smithsonian center. Imagine the waterfront feasts thousands of years ago, with Native Americans cracking open meaty 10-inch blue crabs, instead of the measly 5-inchers common in restaurants today. Why were crabs so much bigger back then? Because intensive fishing pressure today removes at least half of all the existing blue crabs every year—with the largest ones taken first, and the smallest saved. Crabs don't live nearly as long anymore, and so they grow only half as large.

After the English colonists arrived, American shad and oysters assumed favored positions in their diet. Crabs didn't climb on top until the 1930s, with the invention of the box-shaped trap called the crab pot. This device allowed the seafood industry to start catching blue crabs in volumes large enough that they could be widely marketed and sold. The annual catches of blue crabs varied from year to year, but they remained relatively high from the 1940s through the 1980s. However, hidden beneath this seeming constancy was an increase in fishing effort, with the number of crab pots in the bay multiplying more than 10-fold from 60,000 in 1948 to 665,000 four decades later.[3] In the 1990s and 2000s, continued overharvesting of crabs took a serious toll, combined with water pollution and declining water clarity that killed off vegetation that baby crabs need as habitat. Because of the decline, today if you buy a "Chesapeake" blue crab cake in a local restaurant, very likely you are unknowingly eating an entirely different species: the Asian blue swimmer crab, *Portunas pelagicus*, imported from the Philippines, Indonesia, or Thailand. And even if you get your claws on an actual blue crab in a Maryland eatery, it may well have been trucked in from Texas, Louisiana, or North Carolina.

The population of blue crabs in the Chesapeake Bay is erratic and difficult to manage. This is because the life cycle of the species is highly complex and dependent on the whims of wind and weather. However, under the right conditions, crabs can multiply quickly, because they mature rapidly (in as little as a year) and because each female crab can produce as many as three million eggs. So a single good reproduction year can quickly multiply the bay's blue crab population.

The scientific name for blue crabs, *Callinectes sapidus*, is Latin for "beautiful savory swimmer." In reality, however, the movement of the swimmers is a bit *unsavory*—or at least unorthodox—in that they rocket sideways through the water, propelled by their swim fins, instead of the frontward direction that is more normal. The males ("jimmies") and females ("sooks") are different in appearance—with the males boasting, on their undersides, a narrow pencil-shaped strip (called an "apron"), and the females displaying a wide, dome-shaped flap. During the spring and summer, after the females molt out of their hard shells, the jimmies mount the soft-shelled sooks. The males remain doubled up with their mates—inseminating them while guarding the females from competing males and also predators (including other crabs) until the shells of the females have hardened up. Then the males swim off to spend most of their lives in shallow tributaries, while the females head to deeper waters.

In the fall, the females—storing the sperm inside their bodies—begin a long migration southward into the bay's deepest places, near the mouth of the Chesapeake at the Atlantic Ocean. This annual migration is called the "march of the sooks." It's not as militaristic as it sounds, however—with the females meandering in a ragtag parade across the bay bottom, or swimming sideways through the waves, as they make their way from Maryland and other more northern waters to the southern end of the bay in Virginia. The females bury themselves in the sediment at the bottom of the bay during the cold winter months. When the water warms, the females produce fertilized egg masses that look like golden sponges. The eggs hatch to produce millions of larvae, called zoea.[4] These microscopic offspring float out into the Atlantic Ocean, drifting over the Continental Shelf. After about 30 to 40 days, they grow into what look like tiny transparent lobsters, each about 1 millimeter long. These megalopae, as they are called, swim and are swept by tidal currents back into the Chesapeake Bay, as they propel themselves up into the water column on nighttime flooding tides and remain on the bottom at other times. If the winds and currents don't flow in the right direction in the right season—or if storms roll in from an unfavorable direction—the megalopae remain stranded in the ocean and die. They are a form of plankton—fish food—and are often gobbled up by predators. In the years immediately following bad weather conditions, the bay often suffers from low crab populations.

When the winds and waves and rains are right, healthy numbers of the megalopae move from the Atlantic into the Chesapeake. Inside the bay, the megalopae seek shelter from predators in forests of eelgrass in the southern bay. Hiding in the soft and swaying strands, the megalopae develop into juvenile crabs, which shed their shells several times as they grow. This protective habitat of aquatic vegetation has been shrinking in recent years, however, in part because global warming has been making the bay too hot for eelgrass in July and August.

The relationship between blue crabs and pollution is as murky as the bay itself. One might think that spikes in pollution would cause immediate drop-offs in beautiful swimmers. Not true. A fascinating example of this came in 2011. That year, the Chesapeake Bay had some of the muddiest waters on record because huge storms flushed vast amounts of sediment from Pennsylvania farms and elsewhere into the estuary. The bay's health sank to a miserable D+, as measured by the University of Maryland's annual report card.[5] At the same time, however, the number of blue crabs estimated in the bay rose nearly 70 percent compared to the previous year, from 452 million in 2011 to 765 million in 2012, according to winter dredge surveys by the Maryland Department of Natural Resources.[6]

At a press conference in April 2012, Governor Martin O'Malley proudly released the results of that year's survey, which he said demonstrated that blue crab populations had rebounded after crashing to an all-time low of 251 million in 2007. The numbers were impressive: crab populations had *tripled* over five years. That appeared to be evidence of a dramatic comeback for a fishery that was declared a disaster by the federal government in 2008, allowing watermen to collect disaster relief payments.

"The population of the blue crab is now at a 19 year high," O'Malley declared at the 2012 press conference. "A round of applause, please!" he implored the crowd.

This certainly *was* good news for blue crabs. But it was also a challenge to logic. Why did the crabs do better when pollution got worse?

After the press conference, I posed that question to Thomas Miller, director of the Chesapeake Biological Lab. He explained that water quality doesn't have as much impact on crabs as people think. The 2008–12 rebound of the species was caused largely—but not entirely—by restrictions on catching female crabs imposed by Virginia and Maryland in 2008. The protection of mother crabs, and especially Virginia's ban on dredging hibernating pregnant crabs from the mud during the winter, allowed many more crabs to reproduce.

"The blue crab was really affected most by overfishing," Miller explained. "A decade ago, crabs were experiencing 70 percent removal rates in many years. If you take 70 percent of the crabs, you no longer have a healthy crab stock."

But I persisted in my questioning: how could pollution be irrelevant to the blue crabs? For decades, we've heard how important it is to reduce pollution in the bay. And yet, it seems the Chesapeake's best-known tenants could not care less what we dump on them.

That's an overstatement, but it contains a grain of truth. Crabs can often dance around the low-oxygen "dead zones" caused by nitrogen and phosphorus pollution because they are highly mobile.

"Blue crabs are an extremely resilient species," said Lynn Fegley, then Maryland's deputy state fisheries director. "I've seen crabs survive in a bilge for quite a while. And I'm certainly not comparing the Chesapeake Bay to a bilge. But crabs are tolerant."

Fegley posed the following question: how many more crabs would there be today in the Chesapeake Bay if the water were cleaner? The answer is a lot more, because there is an *indirect* but important connection between pollution and crab populations. Low-oxygen "dead zones," created by algal blooms that thrive on nitrogen and phosphorus pollution, kill the clams and worms that crabs eat. Sediment and algal blooms block light that underwater grasses need to survive, and as mentioned earlier, young crabs need the grasses as shelter to avoid predators. In other words, bad water quality limits the amount of crab food and shelter in the bay, which puts a cap on the overall population. Over the decades, pollution has been slowly but steadily pushing that cap downward.

After the impressive crab rebound from 2008 to 2012, populations in the bay then fell more than half—tumbling from 765 million in 2012 to about 300 million in 2013, according to the Maryland Department of Natural Resources.[7] Why? It was not because there was a sudden influx of pollution that year. A more likely—but strange—explanation was that there was a multiplication of large predatory fish called red drum that year, which ate the crabs. The red drum boom in the lower bay, some scientists theorize, may have been triggered by unusual storm winds from the East which blew large numbers of the juvenile fish from the Atlantic Ocean into the sheltered marshes and tidal streams of North Carolina, allowing more to survive predators.

The crab numbers should have rebounded again after this freakish event, but they instead headed downward again in 2014, this time perhaps because freezing weather conditions killed many crabs. In 2015, the estimated crab population bounced back up to 411 million, and in 2016, it rose to 553 million, before falling off again to 455 million in 2017.[8] Biologists guessed that the increase in 2015 and 2016 may have been caused by milder temperatures and more favorable winds. But that's speculation. It may well have been rainfall, or a decline in other predatory fish—or perhaps a waterman's prayer. "The crab stock has been on a rollercoaster ride for most of the last decade," said John Bull, commissioner of the Virginia Marine Resources Commission (VMRC).[9]

A veteran crabber I met on Tangier Island was dismissive of the so-called crab "experts" at the VMRC and elsewhere. "A lot of them don't even know the water is wet," declared Leon McMann, 85, who has been crabbing since World War II.[10] "People in high places don't know nothing about crabs. And we don't know nothing about 'em, neither. All we know is they come, they go, and they bite. That's about all anyone knows about the crabs."

That's a great quote, and I feel like ending my chapter there, but his statement is not really true. When it comes to blue crab populations, there *are* two factors that we know we can control: how many crabs we pull out of the bay, and how much pollution we put into it. These factors clearly matter. But they are also influenced by the whims of nature—which, as it turns out, increasingly isn't all that natural, with greenhouse gas pollution twisting our temperatures, rains, and winds in unknowable ways.

STRIPED BASS

Recovery and Sickness in Maryland's State Fish

A S THE SUN SET and seagulls wheeled around our boat, Paul Spitzer caught something big at the mouth of the Choptank River on Maryland's Eastern Shore. "This is a nice one!" he shouted. "Come on!"

His friend Jim Price, a retired waterman, scrambled for his net. "I'll get him off, right quick."

Spitzer hauled in a striped bass, long as my arm. It was the color of steel, with seven black stripes running the length of its white belly—and something wrong with it.

"This is a striper with lesions," Price said.[1] "The fish could be diseased. It could have *myco*. I could tell you once I cut it open, but it's obviously a troubled fish."

Striped bass are the most popular sport fish on the East Coast. But there are a lot of them like this one—with ugly sores on its sides. When Price cut it open, we saw masses of gray, ball bearing–sized lumps in its spleen, which are the signs of a chronic wasting disease called mycobacteriosis.

The next day, I called a leading researcher of mycobacteriosis in striped bass, Wolfgang Vogelbein, a professor at the Virginia Institute of Marine Science. I asked him how common it is to find masses of these growths inside Maryland's state fish.

"Well, in the Chesapeake Bay here, it is extremely common," Vogelbein replied. "Some of our data over the last 10 years suggests that the 4 to 6 year old fish—the schooling, resident striped bass here—are showing the disease in epidemic proportions. Over 90 percent of those animals are infected."

And yet, oddly enough, striped bass remain relatively abundant. This suggests that the wasting disease, which is caused by a bacterial infection related to the one that causes tuberculosis in humans, is having at most a modest impact on the survival of the fish, whose populations have declined about 20 percent over the past decade. The disease does not infect people who eat the fish. But it does appear to be making stripers scrawnier, with a shorter life span. However, mycobacteriosis is not having as major an impact on the fish population as overfishing did in the 1970s and 1980s, as well as earlier in the twentieth century. And its origins remain a puzzle.

Striped bass are called "rock fish" because they tend to congregate in eddies behind rocks in the rivers where they spawn. Author Dick Russell, who wrote a book titled *Striper Wars: An American Fish Story*, argues that stripers (*Morone saxatilis*) should also be considered "the aquatic equivalent of the American bald eagle" because of their size, power, and deeply ingrained place in American history and culture.[2]

Striped bass range from Canada to Florida along the East Coast. But between 70 and 90 percent breed in Chesapeake Bay, in the freshwater reaches of tidal rivers that feed into the estuary. After the three to five years it takes for them to mature enough to become reproductively active, the females leave the bay and spend the rest of their lives (up to 30 years) migrating up and down the coast, returning every spring to their rivers of origin to spawn.[3] Many males spend their whole lives in the bay and its tributaries. Females tend to spend the summer months in the ocean off the northeastern United States and Canada and the winter months off the Carolinas and Florida.

The reproduction of stripers is a rowdy and bawdy affair. In April or May, when water temperatures warm to about 65 degrees, the females return from their journeys in the Atlantic and exude a scent, which attracts mobs of male bass. Dozens of the males swarm a single female, churning up the water with their tails in a violent splashing called a "rock fight." Pressed by the males, each female releases between 10,000 and five million eggs that look like tiny green balls slicked in oil. The males swim about in the floating clouds of eggs, releasing their sperm (milt) to fertilize them. After the fish leave, the eggs hatch and the larvae grow into fry and then fingerlings. As they grow, these baby rockfish eat water fleas, shrimp, and eventually anchovies and other small fish. Striped bass can grow up to 6 feet long and 125 pounds over a lifetime of about 30 years.

Although Native Americans fished for striped bass, the fish were still so abundant when the colonists arrived that some of the English began to use them as fertilizer on farms. But because the stripers were seen as more valuable as food for human consumption, America's very first conservation law—passed by the Massachusetts Bay Colony, in 1639—prohibited the use of striped bass as fertilizer. In 1776, at the height of the Revolutionary War, when the legislatures in Massachusetts and New York were consumed with such issues as how to fund an army to fight the British, lawmakers still found the management of striped bass so important that they passed bills outlawing the commercial sale of striped bass in the winter. The idea was to avoid overfishing in the winter so that enough females would remain to spawn in the spring. In 1879, striped bass were part of the great American westward migration. The fish were packed into boxcars and shipped on the transcontinental railroad from New Jersey to California, where they were released into the Pacific Ocean and now breed and migrate up and down the West Coast. Although native to the US East Coast, stripers are so popular as sport fish that they have been introduced into waters around the world—including in Mexico, Iran, Russia, South Africa, Ecuador, and Turkey, among other countries. This is a terrible idea from an ecological standpoint, because native Chesapeake species become invasive species overseas. In California, for example, stripers are considered a threat to native, endangered salmon, which the bass gobble up.

The first signs of overfishing of striped bass along the American East Coast came in the late nineteenth century, when Robert B. Roosevelt, the uncle and role model of President Theodore Roosevelt, warned that excessive demand for the fish from the New York market was driving down bass populations.[4] Because of overfishing, stripers nearly disappeared between 1910 and the 1930s. Then they enjoyed a slight rebound when engineers widened the Chesapeake and Delaware Canal, which allowed more water flow into the bass's breeding ground in the Chesapeake Bay. In the 1960s and 1970s, stripers took a hit again, when the popularization of radar and sonar on fishing boats with powerful engines meant that fishermen could catch increasing numbers of the fish. Populations of stripers hit an all-time low in the early 1980s. To prevent a total collapse, Maryland governor Harry Hughes's administration imposed a ban on fishing for striped bass in 1985, and Virginia followed by imposing its own restrictions. The moratorium, which lasted until 1990, was phenomenally successful, as the number of rockfish rose from a low of 8 million in 1982 to 67 million in 2004. The comeback was a boon both to commercial watermen—who fought bitterly against the restrictions—and for recreational anglers, who advocated for the moratorium. The successful cooperation between state and federal agencies to save stripers was perhaps the greatest single victory achieved in the three-decade-plus history of the Chesapeake Bay restoration effort. Since 2004, however, the population of stripers—while still fairly robust—has begun to slowly creep downward again, with the estimated number of spawning-age females declining by about 20 percent over the past decade.[5]

According to Vogelbein, one likely cause of the recent decline is the widespread infection of stripers with this apparently new disease, mycobacteriosis. Vogelbein discovered "myco" in striped bass back in 1997. Over the following 15 years, he devoted countless hours to studying the disease and published several studies on it.[6] Vogelbein concluded that while the bacteria are harmless to people who consume the fish, the organisms cause a slow-acting wasting disease in stripers which makes adult striped bass skinnier and increases the death rate of older fish by 15 to 20 percent per year.

Where does the disease come from? That's a strange tale. Vogelbein and his colleagues examined the water and sediment—and even other species of fish—and found the bacteria almost everywhere they looked. The researchers concluded that it is possible that myco has been in the bay for hundreds of years or longer, with a form of the disease reported in carp as far back as 1897.

This led Vogelbein to the hypothesis that many stripers are becoming sick from this bacteria only recently because environmental problems in the Chesapeake have been weakening their immune systems. The fish can no longer fight infections because they are being stressed by pollution, low oxygen levels, rising water temperatures, and not enough food.

Small fish called menhaden—the primary food for striped bass—have been overfished by industrial fleets out of Virginia. So perhaps protecting menhaden could help stop the disease. Or, maybe not. Vogelbein and a colleague at Old Dominion University, David Gauthier, speculated that menhaden could actually be *giving* the disease to the striped bass.

"We've actually looked at menhaden and it turns out that menhaden harbor a tremendous number of these bacteria in their tissues. And yet they don't develop the disease," Vogelbein said. "And so there is this thought that maybe these animals could be the vectors of the infection to the striped bass. Because at this point we really don't even know how the striped bass get infected."

This would seem to be a critical question for the future of two of the East Coast's most important fish species: striped bass and menhaden. But, sadly, Vogelbein's federal funding for this research ran out recently during a time of budget cuts.[7] That may be a symptom of a disease frighteningly common in humans: shortsightedness.

AMERICAN EELS

*River of Trouble for a
Backward Fish*

I RVING CHAPPELEAR is a Chesapeake Bay eelman. His father taught him long ago that April is the month when masses of American eels swarm around their dock in southern Maryland. That's when it's time to set the traps.

So on a cold and drizzly morning, he set off in a motorboat across the lead-gray Patuxent River.[1] He gripped the wheel with his left hand, while with his right he swung a steel hook on a pole. He snagged a buoy and then hauled up the line until a wire basket broke the surface. The trap was oozing slime and teeming with a knot of greenish eels.

But Irving's next two traps were empty. As the rain started to fall, he threw the empty traps back into the water and wondered out loud, "Where have all the eels gone?"

American eels (*Anguilla rostrata*) are bizarre creatures, and the reasons for their steep decline along the Atlantic Coast are as mysterious as they are. Eels are the most backward fish in the sea: the only species that breeds in salt water and then migrates inland to freshwater streams to live.

Although no one has ever seen it, all of the American and European eels are believed to spawn in a single place—deep in the Atlantic Ocean, east of Bermuda, under thick rafts of seaweed that blanket the Sargasso Sea. Their larvae, called leptocephali, morph into transparent leaf-like creatures called "glass eels," which are then swept by the gulf stream toward the coasts. As they drift toward shore, these glass eels change colors, turning into yellow-tinted "elvers," which become green and then black eels. The creatures spend most of their lives far inland, in ponds and creeks, lurking under rocks during the day and hunting for bugs at night.

Eels are slow to mature, growing for 8–24 years in streams before

they are ready to reproduce. When that finally happens, their eyes bulge
to twice their normal size, and their skin turns silver. Driven by lust, they
thrash thousands of miles back downstream to the spot in the ocean where
they were born, without eating the whole trip. Instead, they digest their
own stomachs for energy. At the end of their epic journey, amid the seaweed and rotting shipwrecks in the Sargasso Sea, the silver eels release
millions of eggs and die, their death spawning new life.

Because their life cycle is so dependent on ocean currents, one theory
is that eels have been declining over the past 30 years because climate
change is slowing their conveyor belt. Disruption of weather patterns
could explain why eels have virtually disappeared from Canada and Europe, but not yet Chesapeake Bay, which is closer to the Sargasso Sea.

But scientists say that there are other factors in their decline, too, including overfishing. In Maryland and other Atlantic states, for example,
a regional fisheries management board called the Atlantic States Marine
Fisheries Commission (ASMFC) concluded in 2012 that the population
of American eels is "depleted."[2] The commission wrote that this is "likely
due to a combination of fishing pressure, habitat loss due to river/stream
blockages, mortality from passing through hydroelectric turbines, pollution, disease, and unexplained factors at sea," with this last item being a
reference to the changes in ocean currents. Harvests of eels up and down
the East Coast have plummeted by more than two-thirds since the early
1980s, tumbling from about 3 million pounds annually to 884,000 pounds
in 2015, with the most severe declines in the northern and southern ends
of the range, according to ASMFC data.[3] Because of the decline, in 2014
the International Union for the Conservation of Nature put American eels
on its "red list" as endangered species.

To help protect the species from overfishing, most Atlantic states
tightened up their regulations in 2014 by increasing the minimum size
limit for catching eels from 6 inches to 9 inches. And yet, the commercial
harvest of eels in the Chesapeake Bay continues to rise, especially for
export to markets and restaurants in Europe and Asia, but also for use
as bait by crabbers. From 1989 to 2009, the annual average harvest in
Maryland was about 300,000 pounds of eels. But from 2010 to 2015, it
nearly doubled to 577,000 pounds a year, according to the Maryland Department of Natural Resources.[4] Biologists believe that this trend is being
driven more by overfishing fueled by the rising prices that watermen can
get by selling eels than by any increase in the eel population.[5] "When the

price of eels goes up, more watermen put out eel pots," said Steve Mink-kinen, a biologist with the US Fish and Wildlife Service.[6]

Few people in America eat eels, outside of sushi restaurants. But eels—especially glass eels—are high-priced delicacies in European and Asian restaurants and markets. In London, jellied eels are popular, served with lemon, onions, and chili pepper. In Dutch kitchens, eels are smoked, shredded, and served with red cabbage.

Beyond overfishing, other problems sinking American eels are industrial chemicals called PCBs, Japanese parasites that bore tiny holes in the swim bladders of eels, and dams blocking rivers. More than 400 dams block streams and rivers in Maryland alone, and about 80,000 dams block waterways across the United States. Dams pose two problems for American eels: First, dams prevent eels from migrating from the ocean where they bred to the freshwater streams where they spend most of their lives. Second, oddly enough, dams alter the sexuality of eels. When they are packed too closely together at the base of dams or elsewhere, young eels tend to develop male sexual organs. When they are able to separate and swim freely, a normal number of eels develop female sexual organs.

On this rainy morning, Irving Chappelear hauled in 40 traps from the Patuxent River and collected a few dozen eels—enough to pay for his fuel, but not enough to earn a good living.

He opened his eel trap and dumped his modest catch into a plastic barrel. One of the slimy snake-like critters flipped away and squirmed across the deck.

He picked it up, dropped it, chased it, grabbed it again, wrestled with it, dropped it a second time, and finally manhandled it into the tub. "In the winter, the slime isn't so bad," he said, wiping his hands. "But in the summer, it's so sticky it's like glue."

As he motored back to shore, the eels wrestled in the barrel, their bodies making sucking, squeaking sounds as if they were crying in a strange tongue. I imagined that they were expressing anxiety about the loss of their world: their vanished siblings, their blocked rivers, and the end of their great journey.

STURGEON

The Dinosaur Matchmaker Meets His Match

A FISHERMAN checking his nets off Smith Island in the Chesapeake Bay felt something heavy. When he hauled it in, he saw that it was, in fact, Jurassic: 6.5 feet long, with five rows of bony plates armoring its sides, a toothless mouth, whiskers, and sandpaper hide. It was an Atlantic sturgeon—one of the few survivors of a 200-million-year-old species that many thought had disappeared from the bay.

As soon as the fisherman motored back to shore with his catch, he called the Maryland Department of Natural Resources, which was offering $100 cash rewards for live sturgeon that watermen caught unintentionally.[1] A team of scientists sped over in a pickup truck, and with considerable effort, they wrestled the beast into a tank and drove it to the lab of Andy Lazur.[2]

Lazur, a biologist at the University of Maryland Center for Environmental Science, is a matchmaker for dinosaur fish, a sturgeon fertility doctor. His dream is to breed Atlantic sturgeon and study their possible reintroduction into the bay. Finally, he had a client: a male for whom he could seek a romantic partner.

Lazur invited me and others to his lab as witnesses to this historic moment, and we watched as the truck rolled up. Four men struggled to lift the sturgeon with straps and slide it into a larger tank in the lab. But the fish slammed its tail, rocking the truck, spraying water, and jolting the men backward, drenched.

As he watched the battle, Lazur admitted that he was excited. "You can't help but have your heart skip a few beats," Lazur said. "It's a species that humans nearly wiped out more than a century ago. Hopefully, we now have the capacity, as good stewards of the bay, to bring this animal back."

Later in his lab, Lazur carefully extracted sperm from the fish and froze the cells. Then he let the sturgeon go. All he needed was a mature female. But then years went by, and no more Atlantic sturgeon were reported by watermen. Finally, another fisherman called to say that he'd netted a big one in the Choptank River—7.5 feet long, 170 pounds. It was almost like a miracle to Lazur, because this time it was a female, full of eggs—the first mature female caught in the Chesapeake in almost three decades.

Lazur brought the new fish to his lab and then started a delicate fertilization procedure that he hoped would bring new life to a legendary species.

Sturgeon are passive giants that were once the kings of the Chesapeake and other waterways around the world. Although almost extinct now, once they were so common that during spring spawning runs thousands of sturgeon—sometimes 800 pounds and 15 feet long—would crowd streams like living logjams.[3] Native Americans told stories about scrambling onto the backs of sturgeon and riding them like broncos.

Atlantic sturgeon (*Acipenser oxyrinchus oxyrinchus*) spawn in rivers with rocky bottoms. They then spend much of their lives—which sometimes span six decades—in the Atlantic Ocean, migrating up and down the coast. Sturgeon were among the largest inhabitants not only of the Chesapeake Bay but also of the Hudson and Delaware Rivers, among many other waterways. The Delaware was once so packed with sturgeon that Philadelphia ferry passengers occasionally saw the monsters leap out of the water and crash down into their boats, soaking the people. In the Chesapeake Bay, tales surfaced of a sea monster with a ridged back known as "Chessie," which could have been a large sturgeon. Similar stories about hoary, ridged-back sea beasts arose in Loch Ness (Nessie), Lake Champlain (Champ), and British Columbia's Lake Okanagan (Ogopogo). All of them could have been the same strange and ancient species.

When I was growing up, I spent my summers near Sturgeon Beach on Lake Michigan, but I never saw a sturgeon, dead or alive. So, as a child, I assumed they had long ago vanished, like plesiosaurs. My father told a story about how, when he was a child back in the 1940s, he was walking along the beach one day and found a large, bony-plated sturgeon washed up on the sand. Every time I walked down that beach, I wanted to find my own sturgeon. When I went sailing, I would peer down over the side of the boat, into the shallows, always hoping to see a ridged-back monster flying along beneath my hull. But that seemed about as likely as spotting a unicorn. For me and others, sturgeon became mystical symbols of a lost America that existed before Europeans arrived. To dream about the return of sturgeon is to imagine a future in which our civilization's environmental crimes are washed away and forgiven.

Sturgeon were a mainstay of the diets of many Native Americans. They speared the fish from canoes and used every part of the animal. The English, however, regarded sturgeon mostly as a trash fish, with their meat fit for savages and slaves. The bony plates of sturgeon made them a nuisance to fishermen, because the plates ripped up fishing nets. In some places, when sturgeon were caught, fishermen would just heap them up on shore and burn them like cordwood. Other sturgeon were speared and sliced apart for their swim bladders—called isinglass—which was used as a clarifying agent in the brewing of beer. Tastes changed in the 1870s, and sturgeon eggs—also known as caviar—became a salty delicacy in Europe. The market for sturgeon soared. They were slaughtered by the millions, with only their eggs extracted, and the rest of the fish left to rot. Like pregnant sea turtles crawling ashore to lay their eggs, sturgeon were especially vulnerable during their breeding season. As their heavy, clumsy, primitive bodies crowded up streams in the springtime to lay their eggs, these toothless giants were all too easy to spear or hook. Sturgeon were massacred almost out of existence in the mid- to late nineteenth century, and competition for the few survivors drove prices for their eggs even higher. Caviar became so valuable that it was like black pearls—a status symbol, an icon of humanity's attitude toward our fellow animals. Like sperm whales, bison, passenger pigeons, and fur-bearing seals, within a few decades nearly all of the sturgeon in North America were wiped out. By 1900, virtually all caviar production shifted to Russia, which then proceeded to make a bloodbath of the sturgeon in the Caspian Sea. By the mid-twentieth century, just about all caviar around the world had to be produced in fish farms because wild sturgeon were basically wiped out. It seemed that the ancient souls of our waterways had disappeared.

On January 31, 2012, the National Oceanic and Atmospheric Administration listed five populations of Atlantic sturgeon—including in the Chesapeake Bay and New York—as endangered species.[4] This granted the few remaining fish additional protections from being captured or killed. Federal law, since 1998, had already made it illegal to catch or keep Atlantic sturgeon, and many states had been protecting them longer. For example, Virginia outlawed the catching of sturgeon for commercial purposes in 1974, and Maryland and New York followed in 1996. But the federal endangerment designation in 2012 offered a new layer of protections, including requiring modifications to fishing gear to avoid unintentional "bycatch" of sturgeon.

In New York, the ban on catching sturgeon appeared to work, with the number of the fish counted in the Hudson River rising over the following two decades. Scientists with the New York Department of Environmental Conservation recorded an average of nine juvenile sturgeon caught per 10 netting efforts in 2006 and 69 per 10 nets in 2015.[5] These numbers remain low but appear to be inching upward.[6]

Maryland scientists in 2014 caught eight mature adult sturgeon in Marshyhope Creek, a tributary to the Nanticoke River on the Eastern Shore, and then tagged and released the fish.[7] The next year, the researchers caught and tagged seven male sturgeon in the same waterway. Finding the males and females in the same river was a positive sign—although nobody has yet found evidence of reproduction in Maryland waters.

One of the last places sturgeon are known to be still reproducing in the wild in the Chesapeake Bay region is in the James River. Sturgeon played a critical role in the history of the river and Jamestown, the first successful English colony in North America. Three-quarters of Jamestown's 240 settlers died in the harsh winter of 1609–10 during what they called the "starving time." Then in the spring of that year, the survivors saw something miraculous: an armada of dinosaur fish, Atlantic sturgeon, returning up the James River to spawn.

"When they saw this huge influx of sturgeon coming back in, they said they could literally go out in the river and spear these fish with their swords," said Chuck Frederickson, the Lower James Riverkeeper.[8] "It was a huge food source and it probably saved the colony from starvation."

Albert Spells, Virginia fisheries coordinator for the US Fish and Wildlife Service, described sturgeon as the "foundation fish" for the Jamestown colony—because sturgeon allowed the English to maintain their toehold in the New World.

"The sturgeon is the animal that saved America," Spells proclaimed. "But for Atlantic sturgeon, we might be having this conversation in French or Spanish right now. So we stand on the backs of sturgeon."

He and allied researchers at Virginia Commonwealth University (VCU) are trying to help the last few hundred sturgeon in the James River reproduce.[9] In 2010, VCU scientists and partners directed barges to dump tons of granite into the river. That created a football field–sized nursery for the fish, which must lay their eggs on hard, rocky bottoms. Unfortunately, too many river bottoms are now buried in silt from farms and suburban construction sites, which smother sturgeon eggs. The hope is that building more rocky breeding grounds for sturgeon will allow them to someday repopulate the whole Chesapeake Bay and East Coast.

Researchers also attached sonic transmitters to the fins of more than 150 sturgeon they caught in the James. The monitors have sent signals—like sonar from submarines—that suggest that sturgeon are being drawn to the artificial reef, which is a positive sign.

"Five or six years ago, most of the biologists assumed that sturgeon were extirpated—gone from the Chesapeake Bay," said Greg Garman, director of environmental studies at VCU. "We now realize that there is, in fact, a viable population in the James."

The researchers have confirmed that sturgeon are spawning in both the James and York Rivers. Using the tracking devices, scientists traced the movements of the James River sturgeon not only to other Virginia rivers—like the York and Pamunkey—but also to New Jersey and Georgia.

According to Matt Balazik, a researcher at VCU, sturgeon have overcome dire conditions in the past and may yet survive humanity. "You've got to realize that these fish have gone through mega-volcanos, meteorites—the things that killed the dinosaurs," Balazik said.[10]

The road to a comeback, however, could be a long and hard one. This is in part because sturgeon are among the slowest-reproducing fish on Earth, spawning once every two to six years, and requiring 15 years to reach maturity. Safe breeding grounds for sturgeon are also rare, with many rivers choked with mud, fishing nets, and boat propellers. So even under the best conditions, restoring sturgeon populations could take centuries or millennia.

Back in his lab at the University of Maryland, Andy Lazur faced a difficult task in trying to play fertility doctor with the unusual creature in his tank. He removed about 40,000 eggs from the female and mixed them with thawed sturgeon sperm he had been saving to create embryos. At first, the fertilization seemed to work, as a few of the cells started to divide, giving Lazur hope. But then the division halted, and the cells died.

Lazur tried again. He worked to reduce the stress on his female, hoping to coax her to produce more eggs. But in the end, it didn't work—and

Lazur felt compelled to eventually release her, as he had released her mate. And after that matchmaking attempt, Lazur hung it up.

So the dating service for dinosaurs went extinct. Outside the sterile confines of his laboratory, however, the last survivors of the fish that saved America continue to quietly swim and breed in the rocky waters that have been home to sturgeon beyond human memory.

The Policies

A COMPLEX WEB of government policies intersect in the Chesapeake Bay—from fisheries management to transportation planning. This book doesn't attempt to tackle all of them, as such an enterprise would likely consume more than one writer's lifetime. So in the following series of chapters I analyze a few issues that I believe are the most pressing.

At the top of my list is the enforcement of environmental laws. I'll detail disturbing failures in law enforcement by the EPA and Maryland to stop sewage from being secretly dumped by the state's biggest city, and by Pennsylvania to halt its farmers' overapplication of manure, which pollutes the bay.

To explore agricultural issues in more depth, I will introduce you to a pair of farmers—one conventional, the other organic—and their widely divergent methods and views.

Then we'll take a hike along a forested stream in a remote Appalachian valley, where a scientist made a discovery about air pollution that turned the bay upside down.

Next, we'll visit the part of the bay that is most threatened by sea-level rise, Hampton Roads, Virginia, to discuss climate change with a retired admiral at the world's largest naval base.

I'll investigate the trendiest new environmental policy idea—pollution trading—and what it means that the bay's leading advocacy organization has embraced this dubious scheme.

Finally, I'll examine the evolving series of bay cleanup agreements signed by the EPA and the states over the past three decades and analyze how they have redefined victory (and papered over defeat) for the multi-billion-dollar bay restoration effort.

ENFORCEMENT

Words versus Water

S OME ENVIRONMENTAL policy experts have trouble appreciating a simple truth: environmental laws are useless if they are not enforced. What appears to work on paper for the Chesapeake Bay often bears little resemblance to what is actually happening in the water. A stark example of this chasm can be found beneath the streets of Baltimore.

Back in 2002, the US Justice Department sued Baltimore because its leaky sewer system was releasing so much raw human waste into the Inner Harbor and Chesapeake Bay that it was violating the federal Clean Water Act. To settle that lawsuit, then-mayor Martin O'Malley signed a consent decree that required the city to fix the problem and eliminate all sewage overflows by January 1, 2016.[1] The city more than tripled water and sewer bills for city residents, in part to pay for the repairs, and collected more than $1 billion.[2] But by the deadline, the city was only about half done with the work required. Sewage overflows continued at a rate that averaged more than one a day in 2015, with a total of more than 42 million gallons of sewage discharged that year, according to city and state records.[3]

The EPA and Maryland Department of the Environment (MDE) failed in their responsibilities to oversee the massive project and make sure that the city complied with the consent decree and its timeline. To compound the enforcement problem, after the city missed the deadline, the EPA and MDE in 2016 granted the city an extension of another 14 years, until 2030, to stop the sewage overflows.[4] That eliminated any possibility that the Waterfront Partnership of Baltimore, a coalition of harbor-area businesses and community organizations, would achieve its long-promoted goal of making the harbor "fishable and swimmable" by 2020.[5]

Even as late as 2015, Baltimore continued to deliberately pipe tens of millions of gallons of raw sewage mixed with rainwater into the Inner Harbor—the city's historic center and economic heart—from a pair of outfalls on the Jones Falls which the EPA ordered the city to close in 2010, according to a report I researched and wrote on the subject for the nonprofit Environmental Integrity Project with funding from the Abell Foundation.[6] The city's two sewage system relief pipes on the Jones Falls— a waterway that is the main tributary to Baltimore's Inner Harbor— dumped 335 million gallons into the river in 119 incidents from 2011 through 2015, with 97 percent of these overflows not reported to the public as required by state law, according to city and state records.[7] The EPA and MDE chose not to penalize the city for any of these violations. Oddly enough, federal and state authorities did fine the city $1.8 million for other smaller and accidental sewage releases elsewhere in the city between 2002 and 2015,[8] but not for the much bigger intentional releases into the Jones Falls and Inner Harbor.

The result: water quality monitoring showed that the amount of fecal bacteria in the Inner Harbor exceeded safe levels for limited-contact water recreation (like in the kayaks, dragon boats, and paddleboards often used on the harbor) at least one-third of the time, according to city data.[9] *Enterococcus* bacteria levels were as much as 472 times above safe levels on the harbor near Light Street and in the Jones Falls near Lombard Street.[10]

Ed Bouwer, an expert on water quality and chair of the Johns Hopkins University Department of Geography and Environmental Engineering, reviewed the city monitoring data for my report on the sewage overflows. He said that the fecal bacteria counts were so consistently high that Baltimore should start erecting signs along the Inner Harbor warning of a health threat to kayakers and other boaters. "These exceedances are really high, and I don't think anyone is really aware of the problem," Bouwer said.[11] "I've seen kayakers on the water, and they have no idea. It would certainly be prudent [for the city to raise health warning signs] because then people would be able to avoid direct contact with the water."

Meanwhile, despite the city's obvious lack of progress in solving its sewage mess, Baltimore Department of Public Works officials, until late 2015, touted on their website great "successes" in complying with the sewage consent decree. "All construction projects and SSO [sanitary sewage overflow] eliminations have been performed within the deadlines (or revised deadlines)," the website said in December 2015, just days before the city missed the deadline to halt the sewage overflows, on January 1. "Excellent management processes [are] in place."

Note the city's wording: the *water quality* was not excellent, but Baltimore was reassuring regulators and ratepayers that its *management* was excellent. This is a symptom of a common problem with the broader Chesapeake Bay restoration effort, in which officials often praise themselves for their efforts—but don't like to focus on the actual water quality.

So why did Baltimore fail to fix its sewers? After all, the city had 14 years and more than a billion dollars to tackle the problem. To be fair, the city made some progress, including repairing or lining with plastic 163 miles of pipes by the end of 2015, which was 39 percent of the city's goal of 420 miles.[12] Money wasn't the problem. There is no question that Baltimore is a poor city with many competing demands, and that upgrading a more-than-century-old network of 1,400 miles of decaying sewer pipes is a complex and demanding task. But after the city was sued by the US Justice Department in 2002, the city collected more than $1 billion—and this money, by law, could not be spent on anything other than the city's sewer and water system. So Baltimore officials could not claim, for example, that the funds were needed for more pressing needs in its police department or schools. It would be illegal for the city to transfer the money it had collected for its sewage system to other general operations of government.

A bigger problem was bureaucratic squabbling and mismanagement. According to quarterly progress reports that the city filed with the EPA and MDE, city and federal officials spent nearly a decade after signing the consent decree arguing over the scope of the project.[13] Baltimore engineers designed a series of pipe replacements and upgrades to handle 3 inches of rain in a 24-hour period. That would be a fairly modest storm, about how much rain would fall in the worst weather event expected every two years. With climate change, however, storms of this size are happening more frequently. The EPA rejected the city's plans, explaining that they were insufficient to stop the sewage overflows. Instead, the EPA told Baltimore officials that the city must design a more robust sewer system with enough capacity to handle 5 inches of rain in 24 hours.

Dana Cooper, chief of legal and regulatory affairs at the Baltimore Department of Public Works, said, "We turned in our last plans in 2008, and then did not get a response [from EPA] until 2011, at which point they said, 'Go back to the drawing board.'"[14] In the end, the city and fed-

eral agencies eventually compromised on designing the system for about
4 inches of rain, with some modifications.

David Sternberg, a spokesman for the EPA, denied the suggestion that
the EPA delayed the process with its rejection of the city's initial plans or
lengthy review. "The plan was contingent on eight sewershed studies that
were being performed by the city of Baltimore," he said.[15] "Our decision
came just seven months after all the sewershed studies were completed."

Regardless of who was at fault for the delay—the EPA or Baltimore—
it is clear that this decade of back-and-forth discussion was unnecessary
and excessive. Back in 2002, when the EPA, MDE, and Baltimore signed
the consent decree, the parties should have communicated more clearly
the requirements for the upgrade. That would have saved city taxpayers at least nine years of their waterways and homes being flooded with
human waste. Between 2013 and 2015 alone, more than 400 Baltimore
homeowners filed damage complaints with the city because sewage from
the city's overwhelmed system backed up into their basements, often with
volcanic force.[16]

Doris Brightful, a seventy-nine-year-old retired nurse from the Grove
Park neighborhood of Northwest Baltimore, pointed at a manhole cover
in the street as she recalled how a flood of human feces in September 2015
wrecked her basement furniture, hot water heater, and photo albums.[17]
"As heavy as that manhole cover is, the sewage overflow just pushed it
straight up into the air and the water just came," Brightful said. "And it
being a health hazard, we couldn't come down [into the basement] because of the [fecal] materials, you know. We couldn't step in that. Me
being a nurse, I told my husband, 'No, you cannot go down there.' We just
had to wait it out. But by the time we waited it out, everything in our
basement was destroyed."

In the Glen neighborhood of the city, Brenda Johnson, a retired city
public school teacher, suffered two sewage floods in her basement in 2010
and 2011.[18] Sewage erupted like a geyser from her toilet and shower drain
for hours. "It was horrendous, and it took forever for us to get it cleaned
up," she said.

She filed a claim with the city for damages but was denied. "The city
came out and basically said it wasn't their fault," Johnson said.

This problem of Baltimore sewers flooding homes appears to have been
made worse by the city's lack of knowledge of its own pipe system. When
the project started in 2002, Baltimore engineers said they didn't even know

what lay beneath their streets. "One of the reasons we got into this mess is that the system had not been proactively maintained over the long term; it hadn't been mapped," said Cooper of the Baltimore Department of Public Works.[19] "We did not have a good sense of the condition of the pipes. . . . So that mapping, that condition assessment, that inspection—that's what took all of the time."

Years into the project, city officials said they discovered that a 12-foot sewer main leading into the Back River Wastewater Treatment Plant had either sunk about 4 feet into the ground over decades or been built too low about a century earlier. The result was a major pipe misalignment that caused a pool of sewage 4 feet deep which routinely backed up for 10 miles under the city.

Unlike some other old cities, Baltimore has separate sewer and rainwater (stormwater) control pipes. But cracks in the city's long-neglected sewer lines allowed rain to seep into the sewer system and overwhelm a network that was already backed up because of the misalignment at Back River. During rainstorms, the city used outfalls to intentionally release waste into streams and rivers to relieve excess pressure from the system. By December 2015, the city had closed 60 of its 73 sewage outfalls. But the city said it could not close the remaining pipes—including those dumping into the Jones Falls—because this would trigger even more sewage backups into the basements of homeowners. In fact, by closing many of the relief outfalls before first fixing the backup problem at the Back River wastewater plant, the city inadvertently caused more flooding of homes.[20] "We didn't really know the right order to do things in, necessarily," said Cooper of the Baltimore Department of Public Works. "And so when we closed those 60 overflows, that actually increased the number of basement backups that we saw in the city—because the sewage has to go somewhere."

Of course, if the city had been regularly inspecting its sewer system over the years, it would have known about the misalignment problem at the Back River plant and would have been able to fix that problem first, shortly after the consent decree was signed in 2002. Having better information might have allowed the city to close all of the sewage outfalls by the 2016 deadline without flooding homes or continuing its massive (but unreported) dumping of sewage into the Inner Harbor and Chesapeake Bay.

More significantly, if Baltimore had maintained its sewer lines properly in the first place, the federal consent decree would not have been necessary. But like in many older American cities, Baltimore's infrastructure was ignored during much of the twentieth century—as our nation spent obscene amounts of taxpayer money on suburban highways, tax breaks for the wealthy, and military expansion in foreign lands, among other priorities. It was almost as if we decided that our own cities were foreign lands, unworthy of our attention.

PENNSYLVANIA

*The Bay Cleanup Collides with the
Politics of Rural America*

I WAS CRUISING across the rolling farmland of central Pennsylvania. On the seat beside me was a camera, because I was on the hunt to capture an environmental threat: drilling rigs, which had multiplied across the state with the spread of hydraulic fracturing. But I pulled over and climbed out of my car when I happened on a beautiful scene. A stream sparkled next to a meadow and a more-than-century-old farmhouse with an ornate porch. In the stream, about 20 cows were chest deep, drinking and cooling themselves.

I took a picture of it, although I had brought my camera for a different purpose. And then I reflected on an uncomfortable truth: I had captured an even worse environmental problem than the one I had been looking for. Charming old family farms like this, with their cows in the stream, are actually more damaging to the Chesapeake Bay than fracking. Although farmers are often seen as the "good guys," agriculture is by far the largest source of pollution in the Chesapeake Bay—and Pennsylvania farmers, in particular, are the biggest contributor.[1] Pennsylvania is responsible for 44 percent of the nitrogen pollution that is choking the nation's largest estuary, which is far more than any other state and more than twice Maryland's 20 percent.[2] Fifty-six percent of Pennsylvania's nitrogen pollution comes from agriculture, compared to only about 10 percent from sewage treatment plants or industrial sources, according to EPA and state data.[3] Pennsylvania also contributes 32 percent of the sediment clouding the bay, and 65 percent of the state's sediment comes from farms. Pennsylvania is the source of 24 percent of the phosphorus pollution in the bay, and 57 percent of this is from agriculture.[4] There are simple ways to prevent this farm runoff pollution—such as by raising fences along streams to keep cows from defecating into them. Farmers can also plant rows of trees to serve as natural filters, reduce their application of fertilizer, and move crops back from waterways. But few farmers do these things, because they mean slightly diminished crop production or cost money. Government programs pay for 75 percent of farm pollution control projects in Pennsylvania, but they do not always cover the loss of revenue.[5]

The unfortunate reality is this: to really start improving water quality in the Chesapeake Bay, the EPA would need to crack down on Pennsylvania and its poorly managed farms. And while massive, factory-style farms are a problem in the bay watershed, in Pennsylvania most of the pollution comes from average-sized family farms—which makes the issue all the more challenging, politically.[6] As far back as three decades ago, policy experts recognized that any meaningful cleanup of the bay would be impossible without a major effort from the one-third of the watershed that lies in Pennsylvania's farm country.[7] But that effort never came—and the bay restoration effort slammed into a brick wall for that reason.

The bay region states have now arrived at a critical 2017 halfway assessment point in the EPA's "pollution diet" plan for cleaning up the Chesapeake Bay, which has a goal of substantially reducing pollution into the estuary by 2025. These federal pollution limits imposed in 2010 (also called the Bay Total Maximum Daily Load, or TMDL) were hailed as the "last, best chance" to save the Chesapeake, because the EPA was supposed to start penalizing states that did not meet their targets and milestones. At this midpoint in the plan, however, the evidence is clear that Pennsylvania is not on track to meet its goals for reducing nitrogen and phosphorus pollution.[8] In the commonwealth's largest waterway, the Susquehanna River, the most popular sport fish, smallmouth bass, recently suffered a population collapse, and ugly sores often mar the fish.[9] The river is being choked by algal blooms and increasing amounts of phosphorus pollution.[10]

"We acknowledge in our own assessments [of the overall Chesapeake Bay cleanup] that we are behind, and a lot of that—about 80 percent of that gap—belongs to Pennsylvania," said Jeff Corbin, the top Chesapeake Bay advisor to the EPA administrator.[11]

Pennsylvania is the key to the success or failure of the whole bay cleanup. "I'm going to say this as clearly as I can," said Nicholas DiPasquale, director of EPA's Chesapeake Bay Program office, to a May 2016 audience of environmentalists at the Choose Clean Water Conference in Annapolis.[12] "If Pennsylvania does not succeed, we're not going to succeed. It's as simple as that."

In blunt language, Pennsylvania is dumping on its downstream neighbors. Its farms and cities are badly out of compliance with federal and state goals to improve the Chesapeake Bay's health. Using more diplomatic and bureaucratic wording, the EPA had this to say in a June 2016 report: "Pennsylvania will need to place considerably greater emphasis on increasing implementation in the agriculture sector to address nitrogen and phosphorus, and in the urban sector for all three pollutants to meet its Watershed Implementation Plan (WIP) and Chesapeake Bay Total Maximum Daily Load (Bay TMDL) commitments by 2025."[13]

Pennsylvania's failure to control its pollution is a subject that many Maryland elected officials and environmentalists try to avoid discussing. Critics of increased regulation in Maryland often point to Pennsylvania's foot-dragging as an excuse for why Maryland should not impose more rules or taxes to clean up the bay. The argument sounds like this: if Pennsylvania dumps *twice as much* pollution into the bay as we do and *gets away with it*, why should Maryland be forced to do more?

To be clear, Pennsylvania has invested some in the Chesapeake Bay cleanup effort, including a $1.4 billion partial upgrade of more than 100 of its sewage plants over the past three decades. But that is significantly less than Maryland or Virginia invested in similar efforts because Pennsylvania chose to upgrade its wastewater plants to a lower level than the other two states.[14] And the gap in effort extends beyond funding to a lack of basic enforcement of environmental laws in Pennsylvania. Politically, this lack of effort is likely because Pennsylvania elected officials and residents feel they have less at stake because none of their land touches the bay. The tragedy, however, is that this means that Pennsylvania's local waterways, including the great Susquehanna, also suffer, and so do the anglers, kayakers, and local property owners who want clean streams in their own state. For example, at least 40 cities and towns in the section of Pennsylvania which drains into the Chesapeake Bay still have antiquated sewer and stormwater pipes that are combined and so intentionally dump raw human waste into rivers and streams whenever it rains.[15] That's *four times* as many of these primitive sewer systems as in any other state in the bay watershed. Only 4 percent (7 of 189) of the large- to medium-sized municipal sewage treatment plants in Pennsylvania's portion of the bay watershed have state-of-the-art pollution control systems and are meeting the highest standards for limiting nitrogen pollution, according to EPA data.[16] That's compared to 79 percent of wastewater plants in Maryland (53 of 67), where the state imposed a special fee (called the "flush tax") to eventually modernize all of its large sewage plants to the highest levels.[17] Virginia has also established a special fund for sewage plant upgrades and has modernized 44 percent (42 out of 95) of its sewage plants.

Inadequate sewage plants are not even the worst problem for Pennsylvania. Because most of the bay watershed in Pennsylvania is rural, farms produce several times more pollution than sewage plants or industrial outfalls in the state, according to EPA figures. However, Pennsylvania's rural lawmakers have long fiercely resisted the regulation of farms. In 1980, the Pennsylvania General Assembly actually made it *illegal* for state or local officials to require farmers to fence their cows out of streams, although this is widely recognized as a standard practice to stop a significant source of water pollution.[18] In other words, not only have Pennsylvania's elected officials failed to fix Pennsylvania's pollution problems, but they have also outlawed good environmental policy that conflicts with the financial interests of farmers.

More disturbing yet, Pennsylvania has repeatedly refused to enforce the state laws that do exist to control runoff pollution from farms. For more than three decades, Pennsylvania's Clean Streams Law has required all farms that produce or use manure to obtain and follow plans designed to keep the waste from running off into waterways, with a failure to do so punishable by fines up to $500 for the first day of each offense and $100 for each additional day. These manure management plans, for example, limit rates of manure application, especially in winter, and require that manure not be spread within 100 feet of streams. But the rules are full of loopholes and rarely enforced. An estimated 40 percent of the 32,600 farms in Pennsylvania's portion of the Chesapeake Bay watershed still did not have these plans in 2017, decades after they were due, according to the state and EPA.[19] And so far, the Pennsylvania Department of Environmental Protection (DEP) has made little or no effort to start penalizing farmers who fail to have or follow these pollution control plans. County conservation outreach workers have made some "educational" (nonenforcement) visits to farms to explain the virtues of manure management plans, but with only limited success. Some conservation districts, in fact, flatly refuse to enforce the law, because they are not supposed to be the enforcement arm of the state DEP and they think that it's their mission to help farmers, not to act as "big brother" to help government.[20]

Pennsylvania's decision to turn a blind eye to law enforcement for rural residents is in contrast to Maryland. The latter state started mandating fertilizer management plans for farms in 1997 and has since made sure that 99 percent of the state's 5,332 eligible farms have these plans, according to a 2015 report of the MDA.[21] The MDA randomly selected 772 of its farms for audits in 2015 to determine how closely farmers were following their plans, and it concluded that 69 percent were in compliance and 31 percent were not, according to the state agency. (A study that year by a University of Maryland researcher found a lower compliance rate, about 39%.)[22] Maryland issued $5,600 in fines to 16 farmers who failed to have nutrient management plans in 2015, $32,950 in fines to 47 farmers who failed to fix problems raised in state audits that year, and $30,750 in fines to 123 farmers for late or missing reports, according to the MDA.[23] Maryland's pollution control plans for farmers are relatively weak. Until recently, for example, they (or the MDA's failure to enforce them) allowed Eastern Shore poultry farmers to dump three times more manure on their fields than the crops needed for their nutrient content (phosphorus), which has contributed to rising phosphorus pollution levels in several rivers on the shore.[24]

Because Pennsylvania is doing so little to control the runoff from its farms, a massive amount of sediment laced with fertilizers has washed from fields into the Susquehanna River and become stuck in the reservoir behind a dam 12 miles south of the Pennsylvania/Maryland state line. The amount of muck heaped behind the Conowingo Dam is staggering: there is enough sediment to fill 265,000 railcars, which if lined up would stretch more than 4,000 miles.[25] For decades the Conowingo Dam acted as a pollution trap that helped to protect the Chesapeake Bay. But now that the reservoir is 92 percent full, its days as a filter are over—and it has become a pollution time bomb. Whenever major storms pass through, the force of the Susquehanna River scours millions of tons of stored sediment—loaded with other pollutants, including phosphorus and metals—from behind the dam down into the Chesapeake Bay. During Tropical Storm Lee in September 2011, for example, about 6.7 million tons of sediment poured through the dam all at once—which was, in one week-long burst, about six times the average amount of sediment that normally flowed down the river over an entire year.[26] About 10,600 tons of phosphorus also poured through the dam during Tropical Storm Lee—more than twice the amount for an average year.[27]

But the problem with the Conowingo is not really a dam problem—because the dam does not create pollution; it just holds it, temporarily. The true source of the problem is runoff from Pennsylvania's farms, as well as the state's failure to make farmers follow the manure management plans that are required by law. In January 2016, Pennsylvania's secretary of environmental protection, John Quigley, admitted that a fundamental issue is that his state's farmers have failed to develop a "culture of compliance" with the law. In other words, Pennsylvania has allowed rampant breaking of its Clean Streams Law and manure management laws, with no penalties for anyone. Quigley made this admission during the announcement of Governor Tom Wolf's efforts to restart—or "reboot"—the state's moribund Chesapeake Bay cleanup efforts. "We need to take a hard look at inspection and verification activities because they have been the missing piece," Quigley said during the press conference.[28] "We have got to, in the state of Pennsylvania, create a culture of compliance with existing regulatory requirements. The Pennsylvania Clean Streams Law has been on the books for more than a quarter century. . . . And if we don't go back to the basics of government that works, our efforts relative to the bay—to say nothing of local Pennsylvania water quality—will continue to seriously lag."

During a telephone interview three months later, I asked Quigley whether his agency had finally started taking enforcement actions against the approximately 10,000 livestock farms in the bay watershed which still lack the required pollution control plans. For example, I queried, had Pennsylvania sent out letters asking farmers to produce the required plans within a set time period or face a warning and then a fine? Quigley replied: no, the state was not yet prepared to do something like that.[29] First, Quigley said, the state would work with academic researchers to conduct a comprehensive survey of farms, to determine what kind of runoff pollution control practices they are already following.

In other words, more than three decades after Pennsylvania passed a law requiring pollution control efforts by farms—as well as 15 years after

Pennsylvania and the other Chesapeake Bay states failed to meet a 2000 deadline for cleaning up the bay, and five years after the states blew a second deadline in 2010—Pennsylvania is *just now* planning to start collecting information on the subject of farm runoff pollution. But the state is still not yet taking action to stop the pollution, beyond voluntary efforts that have proven inadequate.

I asked Mark O'Neill, spokesman for the Pennsylvania Farm Bureau, why so few farmers follow the law. He speculated that there could be several factors, including a lack of awareness among some farmers of the requirement for the manure management plans. Some farmers might be refusing to comply because creating the plans can cost a farmer about $1,000 to hire a consultant, or time (which means money) if the farmer writes the plan himself. "Making a profit now in farming in Pennsylvania is extremely challenging," O'Neill said.[30] "Farming is a vocation and a way of life, but it is also business as well. And if they cannot make money, year after year, then they are not going to be around very long to pass it on to the next generation."

Not only has Pennsylvania been slow in its education of farmers about the state's legal requirements to follow pollution control plans, but the commonwealth has also been backpedaling on environmental enforcement, in general. Republican governor Tom Corbett, for example, who held office from 2011 to 2015, cut deeply into the budget of the state's environmental agency (DEP) at the same time that he slashed taxes for businesses, reducing corporate payments by hundreds of millions of dollars annually.[31] Corbett also blocked a proposed tax for the natural gas industry which would have brought in millions of dollars to the state. Corbett and his predecessor in the governor's office, Ed Rendell (a Democrat), slashed the DEP especially harshly. Over the past decade, while the average Pennsylvania state agency lost 6 percent of its workforce, the DEP lost 14 percent of its workers.[32] The DEP lost 671 positions from 2009 to 2016, and 441 of these workers were inspectors and environmental permit writers. Astonishingly, the cuts left only *three DEP employees* to oversee the 32,600 livestock and grain farms in the Chesapeake Bay watershed. The staffing levels are now so low, Quigley said, that the agency is only inspecting about 1.8 percent of these farms every year, much lower than the 10 percent annual inspection rate that the EPA has determined should be a minimum. "So the agency's ability to do its job—and perform basic functions—has been severely compromised over the last decade be-cause of budget cuts," Quigley said. It is a historic fact, however, that even before the cuts, the DEP was not even attempting to enforce its manure management plan requirement for farms.

Beyond the question of staff is the issue of government money avail-able to farmers to help them build pollution control projects, such as streamside fences. Assuming that taxpayers should pay for the pollution control projects on private land, Penn State University has estimated that the commonwealth would need to invest about $378 million per year in pollution control projects between 2013 and 2025 to meet the pollution reduction goals in the EPA pollution "diet" for the Chesapeake Bay (the bay TMDL). So far, Pennsylvania has only been paying for about $128 million a year—leaving a gap of about a quarter billion dollars a year. "That gap is one of the principal reasons we are falling short of our pol-lution control goals," Quigley said.

And this quarter-billion-dollar-a-year price tag doesn't even take into account the cost of upgrading the antiquated sewer systems in the more than 40 cities and towns in Pennsylvania that still dump raw human waste into bay tributaries during rains. Replacing these primitive com-bined sewer systems would cost another $18 billion, and the state isn't even contemplating that right now, Quigley said. "There is no question that this country has retreated, since the Reagan era, from significant public investments in many quarters. And our infrastructure has suffered greatly as a result," Quigley said.

In May 2016, Quigley's already-questionable "reboot" strategy for the bay cleanup was thrown further into doubt when he abruptly resigned in an unrelated scandal. He left his job after he sent an e-mail with foul lan-guage to environmental groups he was angry at for not helping him with his efforts to regulate the gas drilling industry in Pennsylvania.

Evan Isaacson, a policy analyst for the Center for Progressive Reform, said that an obvious answer to Pennsylvania's funding gap for bay resto-ration would be for the state to impose a tax on the extraction of natural gas from its large number of drilling and fracking sites. "Every other oil and gas producing state has a severance tax [on oil and gas extraction]," Isaacson said.[33] "I think it would be a reasonable notion that if you are going to be pulling the gas out of the ground anyway, they might as well tax that gas like every other state does and dedicate a portion of the money to restoring the environment."

Why hasn't the state done this? The political background is that the Pennsylvania General Assembly—dominated by Republicans and rural conservatives—has been hostile to the imposition of any taxes or funding for any environmental projects. Mark O'Neill, spokesman for the Pennsylvania Farm Bureau, noted that his organization sued the EPA to try to block the agency's pollution limits and pollution "diet" for the Chesapeake Bay. "There is categorical distrust of the Environmental Protection Agency, with farmers—not only in Pennsylvania, but across the United States," O'Neill said.

Jacquelyn Bonomo, chief operating officer of Penn Future, the state's largest environmental organization, said that there is a politically "toxic atmosphere" in the Pennsylvania General Assembly toward the Chesapeake Bay cleanup.[34] "The legislature is frankly holding clean water hostage and negatively impacting communities that drink and really need it, as well as the industries that rely on clean water," Bonomo said.

The consequences of Pennsylvania's failures are obvious downstream in the Chesapeake Bay. Every time a major storm rolls in, a giant plume of pollutants the color of chocolate milk gushes down from Pennsylvania into the bay. The muddy stain is so large that NASA satellites can see it from outer space. Most of this muck is from Pennsylvania's farms.

So where is the EPA on this? The problem of farm pollution flowing down from Pennsylvania into Maryland and Virginia is clearly an interstate issue that no individual state can resolve. The EPA is the only entity that could possibly step in and save the bay. However, the EPA has little legal authority over farm pollution, because Congress drafted the 1972 federal Clean Water Act with an exemption for most farms (other than large livestock operations). So unless Congress has the political will to update the Clean Water Act (which would be helpful but is politically impossible given the current conservative, anti-regulatory bent of the House and Senate), the only solution available is for the EPA to take action on polluters that it does have legal authority over, which is sewage treatment plants and industries.

Michael Helfrich, the Lower Susquehanna Riverkeeper, noted that modernizing Pennsylvania's sewage plants to bring them on a par with Maryland's was never a part of Pennsylvania's plan to meet the EPA pollution limits for the Chesapeake Bay. Why? Because these upgrades are much more expensive than getting farmers to build fences or plant trees along streams and take other steps to reduce their runoff pollution. Unfortunately, Pennsylvania has failed to do that as well. "We were trying, from the start, to do the cheapest thing first—but unfortunately, we have not seen the results from that," Helfrich said. "We have a legislature that needs to do something to make enforceable rules to require the pollution reductions from agriculture. But if that does not happen, then we are going to have to do the more expensive upgrades in the wastewater treatment plants."

Time is up. Action to reduce farm pollution simply has not happened in Pennsylvania. And so now it is beyond time for the EPA to intervene—with two bay cleanup deadlines missed in 2000 and 2010 and a third, in 2025, now imperiled by Pennsylvania's foot-dragging.

Jon Capacasa, water program director for EPA Region 3 in Philadelphia (which oversees the Chesapeake Bay region), said that the federal agency is starting to take some limited actions to try to push the state. "We won't be satisfied until the job is done," Capacasa said.[35] "We've objected to permits [for expanded sewage treatment plants]; we've withheld funding [temporarily] and conditioned funding. We've taken targeted compliance actions, both in the [animal livestock industry] world as well as the stormwater world. We've exacted penalties as we need to. And we will continue to look for other tools to use to make sure the job gets done."

For example, the EPA in 2014 objected to 14 proposed sewage treatment plant permits in Pennsylvania—including for the Williamsport Sanitary Authority, the City of Sunbury, and the Dover Township Wastewater Treatment Plant—because they would not reduce pollution enough to meet the EPA's pollution "diet" for the bay.[36] The federal agency also that year sent orders to 85 municipalities in Pennsylvania—including Scranton and Wilkes-Barre—requiring improvements to their stormwater management systems.[37]

I asked Jeff Corbin, the EPA's top Chesapeake Bay advisor, whether the agency finally will start requiring Pennsylvania to upgrade its sewage treatment plants to the level that Maryland and Virginia taxpayers are paying for. That would seem fair, because right now Pennsylvania's sewage pollution is pouring downstream into Maryland and eroding that state's progress. A federal crackdown on sewage plants could be used as a financial lever to make Pennsylvania finally start enforcing its Clean Streams Law for farms.

"We haven't flipped the switch and told them that it's time to do that,"

Corbin replied.[38] He suggested that "the ball is in Pennsylvania's court" with regard to the Chesapeake Bay cleanup.

But really, the ball is in the EPA's court. When—if ever—will the EPA really crack down on Pennsylvania for its years of dumping on its downstream neighbors? The fate of the Chesapeake Bay hinges on whether the EPA can do its job.

AIR POLLUTION VERSUS WATER POLLUTION

Cleaning the Water from the Sky

I N A REMOTE VALLEY in the Appalachian Mountains, as the setting sun lit the tops of trees with golden light, Keith Eshleman strode down a ragged logging road to his workplace. Eshleman, a water quality scientist at the University of Maryland Center for Environmental Science (UMCES), hopped onto a mossy rock as he crossed a stream and then forged through nettles until he was a mile deep into a pathless corner of the Savage River State Forest. Finally, he dropped his backpack beside a creek called Black Lick, where a wooden box was mounted atop a steel pipe. He flipped open the box's lid and revealed the digital display of a water level monitoring device.

It was here, in an inch-deep trickle of water shaded by hemlock trees 180 miles from the Chesapeake Bay, that Eshleman made a discovery that turned the bay upside down. Over two decades of monitoring this and several other streams in Western Maryland, Eshleman found that levels of nitrogen—a pollutant that fertilizes algal blooms and dead zones in the bay—plummeted by 50–70 percent from 1995 to 2010. "It was a complete shock," said Eshleman, who earned a PhD in water resources from the Massachusetts Institute of Technology.[1] "You know, when we do things in hydrology and water quality, we expect to see a one or two percent improvement, or maybe a 10 percent improvement or a 10 percent decline. Those are pretty big numbers. So when we see a 50 to 70 percent reduction or something of that magnitude it's absolutely, totally surprising."

The combined effect of the drop in the nitrogen in all of the streams flowing into the Potomac River was an unexpected 35 percent decline in the pollutant entering the bay from the river. This was extraordinary on several levels. The conventional wisdom was that wooded creeks had nitrates in them, but they were *naturally occurring* from the decomposition of leaves caused by bacteria in the soil. The slow leaching of such nitrates was thought to be relatively constant—not something that could suddenly drop off and improve water quality downstream. Eshleman deduced that the huge change must not have come from the land, because the forests surrounding the streams had not changed much from 1995 to 2010.

"We said: 'Oh my gosh! Something is going on here that we really can't explain,'" Eshleman recalled. "And this is what tipped us off that we need to look more closely at what's coming into this system. Because the outputs are clearly decreasing by leaps and bounds."

When he examined the other major input—the air—he figured out that much of the nitrate in the soil around forested streams was not natural. The nutrient had floated from the smokestacks of coal-fired power plants in Ohio and elsewhere in the Midwest and been carried by rain and snow into the mountains of Western Maryland. And now the streams in those mountains—and across the Chesapeake Bay watershed—were becoming cleaner thanks in part to a federal air pollution control law signed in 1990 by President George H. W. Bush which required the installation of pollution control devices on coal-fired power plants.

In other words, the water was being cleaned from the sky, thanks in part to George Bush. Things with the Chesapeake Bay are not what they seem.

In a scientific journal article published in *Atmospheric Environment* in 2016,[2] Eshleman and graduate student Robert Sabo evaluated what their findings for the Potomac River basin might mean if applied across the bay watershed over the three-decades-plus history of the Chesapeake Bay restoration effort. They concluded that while the modernization of sewage treatment plants produced significant reductions in nitrogen pollution in the bay, more recently those improvements had leveled off because that basic work was mostly done in Maryland and Virginia (although not so much in Pennsylvania, which we discussed in the previous chapter). Since then, the bay region states have shifted their efforts to a far thornier problem: runoff pollution from farm fields, suburban parking lots and streets, and urban pavement. This runoff pollution is much harder to control, in part because it doesn't come from pipes, which can be monitored and capped with filtration systems. The 1972 federal Clean Water Act has strong mandates for controlling pollution from pipes but loopholes for runoff pollution. The most controversial part of Eshleman's paper argues that there is no scientific proof that recent state and federal actions to control runoff pollution from the land have worked to reduce the nitrogen that is choking the bay. These actions have included government incentives to farmers to plant trees along streams and programs that encourage the construction of artificial wetlands. Instead, Eshleman said that the lion's share of the improvements that the bay has experienced in recent years have come from more effective regulation of air pollution. This air pollution is more easily controlled than rainwater runoff, because it comes from pipes—the smokestacks of power plants and industries, and the tailpipes of cars and trucks.

"If there is one kind of interesting irony to this entire story it's that it took air quality regulation to improve water quality," Eshleman said. "So think about that. Water quality regulation is what should be working. But frankly, it appears to us—at least if you believe the data we are sharing— that it was really the air quality component that worked and not the water quality side of the house."

Why did this happen? In a more bipartisan era back in 1990, the first President Bush, a Republican, collaborated with Democrats in Congress to approve landmark amendments to the federal Clean Air Act of 1970. These amendments were designed primarily to protect human health by reducing industry-produced sulfur dioxide and nitrogen oxide air pollution that created smog, soot, and acid rain. Toward this end, the law proved successful beyond expectations, with at least 160,000 premature deaths from heart attacks and lung disease prevented annually.[3] Although critics of the bill warned that the "job-killing regulations" would mean "quiet death for businesses across the country," in fact the "job-killer" claims were a myth (as usual).[4] The real cost to industry of the 1990 law turned out to be about one-third of what opponents claimed, with none of the predicted closings of the auto industry, gasoline refineries, hospitals, and other businesses.[5] In fact, by the year 2020, the air pollution law that Bush signed is projected to have had an impressive 30:1 benefit-to-cost ratio, with $65 billion in costs for pollution control equipment compared to $2 trillion in mostly health-related benefits.[6] Again note that the whole rationale for the law was built around the benefits to human health, not stream health. "When President Bush signed that law, he really wasn't interested in—and didn't believe, frankly—that we would see the positive effects here in the water that we are now seeing," Eshleman said. "This was not even part of the equation."

As it turns out, about one-third of the nitrogen pollution smothering the Chesapeake Bay comes from air pollution.[7] Bush's 1990 law gradually required many coal-fired power plants and other industrial polluters to install devices (called "scrubbers") that capture sulfur dioxide gas, as well as other control equipment. The scrubbers not only helped to protect human health but also reduced the acidity of many streams in the northeastern United States. This curtailing of acid rain was expected. But a side effect of the 1990 law was less nitrogen air pollution falling with the rain into waterways—meaning fewer algal blooms and smaller "dead zones" in places like the Chesapeake Bay—because of the installation of a different kind of pollution control equipment for nitrogen oxide (NOx) gases. These devices, which use a process called selective catalytic reduction, inject ammonia into coal plant exhaust to convert NOx into water and nitrogen gas, which is common and harmless.

Eshleman, a registered Democrat, is emphatic about giving credit to Bush for these improvements:

> From a political standpoint, somebody ought to write President Bush a letter. I know he's not in very good health. But he really needs to know, 'You signed a bill, and now you are protecting and restoring *so much* stream and river habitat, even to the point of helping to restore Chesapeake Bay.' It's really amazing. Look at the time lag between the passage of the law and now: 25 years! It really speaks to the fact that we do things, and we *think* they are the right things to do, but if we don't see an impact within a few months or a year, we give up on it. It's like we don't have the attention span needed. And yet the environmental values that many of us share call for things that you can't do in a week or a month or a year. We didn't destroy these ecosystems in that time frame. It took us *hundreds of years* to produce the problems we are now facing. So why do we think we can fix them within a short period of time—within an election cycle, say?

To be sure, it wasn't only President Bush's Clean Air Act amendments of 1990 that helped to clean up the streams and bay. Perhaps most importantly for the Chesapeake, over the past decade Maryland has gradually upgraded most of its large sewage treatment plants to a state-of-the-art level using money from a 2004 "flush tax" imposed by former governor Robert Ehrlich (like Bush, also a Republican). Virginia has also invested billions of dollars to upgrade almost half of its large sewage treatment plants to this enhanced level. Pennsylvania has upgraded its sewage plants to a lower level. The investments in Maryland and Virginia, in particular, have contributed to measurably cleaner water in the Potomac, Patuxent, James, Corsica, and Back Rivers, among others.[8] Also possibly helping water quality was the real estate crash of 2008, which slowed the expansion of suburban sprawl and blacktop across the Chesapeake region.

Other recent state and federal air regulations have also enhanced water quality, as have changes in technology. In 2006, Maryland lawmakers (led by state senator Paul Pinsky of Prince George's County, a Democrat) passed the strongest air pollution control law in America: the Maryland Healthy Air Act. It mandated the construction of devices on the state's largest coal-fired power plants to cut nitrogen oxide air pollution by 75 percent over a decade. Three years later, in 2009, the Obama administration started imposing additional regulations on mercury, nitrogen oxides, and toxic pollutants from power plants, as well as tighter fuel efficiency requirements for cars and trucks. Innovations in gas drilling technology also may have helped. The rise of hydraulic fracturing and horizontal drilling made natural gas much cheaper than coal, and so power plants are increasingly switching to gas to generate electricity, which means less air pollution falling into the bay (although, in Pennsylvania, some streams and drinking water wells have been contaminated by spills and leaks of fracking fluids).[9] "Gas is so much cleaner when you burn it than coal," said William Dennison, vice president of the UMCES, during a phone interview.[10] In addition, the federal and state governments "are just doing a better job at the smokestack level of regulating those pollutants more stringently, and the bay is seeing the benefit of that."

Back in the mountain valley, as evening deepened, Eshleman picked up his backpack and marched back down the logging road, heading toward his car. There was a spring in his step as he reflected on the success story he had uncovered like a lump of gold in a stream, hidden deep in the mountains. "We can only hope there's more to come," he said, passing a twisted and rotting tree trunk that reached upward into the clear night sky.

AGRICULTURE
A Tale of Two Farmers

I T'S A HOT AFTERNOON in Tuscarora, Maryland, and in an open barn longer than a football field, fans whirl to cool off 170 Holstein and Jersey cows. Chuck Fry feeds his animals and then leads a visitor to a pair of giant tanks holding the more prodigious product of dairy cows. "For every gallon of milk I get I am benefitted by three gallons of manure," said Fry, president of the Maryland Farm Bureau.[1] "Now, that's a curse and a blessing. We use that three gallons of manure to grow next year's crops. So we store it and treasure it because it has tremendous value."

But manure also causes tremendous harm to the Chesapeake Bay, with farm runoff the single largest source of pollution in the estuary.[2] And so Maryland, in 2012, imposed regulations to reduce runoff by prohibiting the spreading of manure in the winter, when the ground is frozen and crops can't absorb it. The regulations also require farmers to mix and incorporate manure into the soil of their fields, instead of just spraying it on top, which leads to more runoff.

The pollution control regulations were to take effect on July 1, 2016. But because Fry and others complained to Governor Larry Hogan's administration about the cost—especially to the state's 430 dairy farmers—the state weakened the regulations. "Those regulations would have driven those dairy farmers out of business," Fry argued, explaining that the rules require the construction of manure storage tanks like his, which can cost tens of thousands of dollars.

Environmentalists protested the watering down of the clean water rules, but the farm lobby prevailed. It was a repetition of a pattern that stretches back decades in the Chesapeake region and from coast to coast in the United States. Farmers, more than any other industry, resent government intrusion into their affairs—although they are happy to accept government subsidies for their businesses. In total, about 35 percent of Maryland's 12,834 farms receive government subsidies.[3] These farmers pocketed more than $1.1 billion in taxpayer money from 1995 to 2014, including $462 million in federal corn price subsidies and $204 million for soybean price supports.[4]

Across the Chesapeake Bay watershed, there are about 87,000 farms and 8.5 million acres of cropland, meaning that farms cover almost one-quarter of the land.[5] But despite occupying this relatively small portion of the real estate, farms produce 55 percent of the phosphorus and sediment pollution fouling the bay and 42 percent of the nitrogen, according to the EPA Chesapeake Bay program.[6] Although the number of farms has been declining, the size of the average farm and the total number of acres in crops have been rising over the past half decade.[7] This is partly because federal ethanol fuel mandates have boosted the profitability of growing corn.[8] The poultry industry on the Eastern Shore is also building more factory-scale houses, and the weight of the birds is increasing, with 4 billion pounds of broilers produced on the Delmarva Peninsula in 2015 compared to 3.3 billion pounds in 2005.[9]

But some of the farmers, especially in the dairy industry, are not doing well. They own valuable land but carry heavy debt and earn meager profits. Fry's family, for example, has been farming their land (they own 200 acres but lease and work 1,500 acres) near the Potomac River about an hour northwest of Washington, DC, since 1883. However, Fry said he doesn't know how much longer they can hold on. The fifty-five-year-old said he earned only about $34,000 from the farm in 2015 (about 13% below average for a farm in Maryland).[10] And in other years, he's been hit with losses, in part because of the depressed price of milk, which is caused by federal price controls and overproduction by industrial-scale dairies in other states, including California and Wisconsin. The national milk glut is drowning the small dairy farms of Maryland, half of which have closed since 1993.[11] "I'm praying that in the future, times will get better," Fry said.[12] "But I guarantee you that we cannot sustain this dairy for another year based on the prices we are receiving now. It's just too much. I'm done. I just can't do it."

In the face of such financial pressures, Fry argues that farmers need fewer rules. "What really needs to happen is that the environmental community needs to really listen to the farmer, and hear what they are saying—

and truly hear it, and believe it," Fry said. "Or we are just going to have to go down the path of farmers need to do what they need to do, and regulators need to do what they need to do, and somewhere in the middle there is going to be a clash. We don't want that."

Fry's story resonates with an American archetype: the family farmer fighting the government to defend the land that has defined his family for generations. In an attempt to save their farm, Fry's wife, Paula, and daughter, Gail, opened a side business, Rocky Point Creamery, which turns their milk into ice cream to sell to commuters who speed past. That appealing Hollywood plotline is muddied, however, by the fact that Fry—as president of the Maryland Farm Bureau—also gets paid to head an organization that lobbies aggressively against laws and regulations that would protect public health and clean up the Chesapeake Bay. The Farm Bureau, for example, fights against restrictions on the spraying of pesticides, herbicides, and insecticides. They also lobby against legislation that would limit suburban sprawl, because this would mean that farmers could earn less money when selling their land to developers. In 2012, the farm lobby organized a "tractorcade" (a parade of tractors around the statehouse) to protest legislation proposed by Governor Martin O'Malley's administration which would have *protected farmland* by restricting the construction of large subdivisions with septic systems in rural areas. The downside, to farmers, was that it would have also restricted their potential income from selling their properties to developers. The Maryland Farm Bureau came out strongly against O'Malley's bill, the Sustainable Growth and Agricultural Preservation Act of 2012. Valerie Connelly, executive director of the Maryland Farm Bureau, explained, "You know, farmers don't have 401(k) plans, they don't have retirement systems. Their retirement plan is whether they sell their land."[13] This is not always true, because some farmers do have retirement accounts (some marketed by the Farm Bureau). But the broader point is that helping farmers to sell their land—not necessarily preserving farms—is a goal of the Farm Bureau.

Although he is earnest in his beliefs, Fry holds views on some issues that are at odds with science and the historical record. For example, during a two-hour interview on his farm, Fry contended (among other things) that maintaining trees beside streams is a threat to clean water, because leaves fall in and decay. In his mind, it's better to have pastures beside streams—although cows muddy the banks and defecate into waterways. He and other farmers boast that farmers were "the first environmentalists"—when, in fact, European farmers were the original polluters of the bay, with English colonists forcing African slaves to clear forests and drain the land for tobacco plantations, which caused erosion and sediment pollution.[14] In that way, Fry is not an actor in a Hollywood movie about the family farm, but a player in a real-life American tragedy. Americans of different political tribes increasingly live inside bubbles of their own private "facts" that bear no relationship to the facts of others or historical and scientific facts—making compromise or progress impossible. If we don't share the same reality, how can we improve it?

In Fry's Farm Bureau reality, farmers are doing more than anyone else to clean up the Chesapeake Bay. But they keep getting "hammered, hammered, hammered" (as Fry put it) with more regulations. In the other reality—the real one—agriculture remains less regulated than most other industries, even though farming today is a big business that is highly mechanized and dependent on the widespread use of chemicals, petroleum, and factory-scale operations. And while farms have done a few things to improve water quality, it's only a small portion of what is needed to restore the Chesapeake Bay to health, and agriculture remains by far the largest source of water pollution. At nearly every turn, the farm lobby has slowed or derailed the bay cleanup effort by demanding and receiving exemptions from many of the environmental laws that other industries must follow.

Here's an example of its influence. Since at least the 1980s, the bay region states have promoted a very simple but important strategy to prevent pollution: fencing cattle out of streams, so that cows can't trample and erode the banks and defecate into waterways, fouling the water for others downstream. But because of the resistance of farmers, more than three decades into the bay cleanup effort, not one of the states requires streamside fencing for livestock. More disturbingly, neither the states nor the EPA's Chesapeake Bay Program even try to track how many farmers follow this best management practice. Rough estimates are that only about one-quarter to one-half of farmers across the region fence their cattle out of streams.[15] But it might be lower than that. The figure was only 20 percent in Virginia's Rockingham County, the biggest agricultural area in that state, when the Shenandoah Riverkeeper in 2016 used detailed aerial photographs from Google Earth to examine the 841 farms with livestock in the county that also have streams or rivers flowing by their pastures.[16]

"I hear the complaints of the farmers," Fry said. "I can think of several farmers who just go on and on about the cost of the fencing. They say if you make them fence the cows out of the creek, they are going to quit. I think farmers are like anybody else. When you tell them they have to do something, their first reaction is: 'No you're not. It's my land, and I pay taxes.'"

Except in this case, nobody told the farmers to do anything. Taxpayer-funded programs in the bay region try to *encourage* farmers by paying 87–100 percent of the cost of raising fences and installing alternative drinking water systems (so farm animals don't have to drink from streams).[17] But despite the subsidies, most farmers still don't follow this basic practice. In fact, in Pennsylvania, the General Assembly in 1980 passed a law *prohibiting* the state's environmental agency from requiring farmers to install streamside fencing.[18]

Fry said he is frustrated that farmers in Pennsylvania do so much less to protect the bay than Maryland farmers. More than 99 percent of Maryland's farms have fertilizer management plans meant to reduce overapplication of manure, which is much better than Pennsylvania's 60 percent.[19] However, a 2015 survey by a University of Maryland researcher discovered that 61 percent of the Delmarva farmers with plans admit that they do not actually follow them.[20]

Fry guessed that some farmers may refuse to follow their fertilizer management plans because they restrict them from applying as much fertilizer as the farmers would like to achieve maximum crop production, which means maximum profit from the land. "Your nutrient management plan is going to restrict you to 150 bushels of corn [per acre], but you know you have the farm potential of 300. Why would you do that?" Fry asked. "It's kind of like buying a Maserati [sports car] and putting a governor on it so it'll only go 55 miles an hour."

Fry grows corn to feed his cattle, and he plants genetically modified seeds that are soaked in an insecticide called neonicotinoids. "Neonics," as they are called, have been linked to a decline of bees and other pollinators around the world.[21] Fry said he also sprays the weed killer glyphosate, or Roundup, on his crops. In 2015, the World Health Organization listed glyphosate as a "probable carcinogen" for humans.[22] Some researchers have also concluded that glyphosate is playing a role in the die-offs of more than 80 percent of North America's monarch butterflies.[23]

Fry dismissed the risks of cancer or ecological damage from pesticides. "If you feed a person enough ice cream, it would probably give them cancer," he said.

The Maryland Farm Bureau in 2016 lobbied against a bill that would have restricted the sale of "neonics"—not because the legislation would have hurt farmers (it only applied to homeowners and their gardens), but because the farm lobby worried that the rules might *eventually* be expanded to include agriculture.

This is part of a bigger picture. Farm advocates argue fiercely against regulations of all kinds, often with the slogan and bumper sticker "No Farms, No Food." Translation: hands off, or children will starve. But the truth is that obesity and diabetes—fed by US government subsidies of corn for soda and junk food—are far bigger public health problems around the world today than starvation, according to the World Health Organization.[24] And there is another way to grow healthy food which does not require chemical pesticides or Chuck Fry's giant tanks of liquid manure.

To look at the alternative, I drove to southern Maryland to visit an organic farm, Brett Grohsgal's Even' Star Farm in St. Mary's County.

At the end of a long dusty road, Grohsgal's place does not look like a modern farm. It looks like a farm from a century ago, with its weather-beaten barns and workers bent in the fields. Instead of having all of his land in one or two crops—corn or soybeans—Grohsgal's organic farm grows 70 different species of vegetables, fruits, flowers, and herbs.

"Farms 100 years ago were really diversified, because you had to be diversified for risk management," said Grohsgal, a fifty-six-year-old energetic former chef with a sunburned face and sweaty blue T-shirt. "Why? Because you didn't have federally subsidized crop insurance. You didn't have the whole mantra of, 'if you don't specialize, you're dead.' You didn't have the whole nonsense of 'Get big or get out.' Farms had to be diversified back then, because they were providing most of the food that nearby small towns ate."

Grohsgal raises his crops on only one-quarter of his 100 acres, with half of the remaining land kept in forests and with another one-quarter in hay. His vegetables—onions, tomatoes, cucumbers, squash, okra, basil, eggplants, sweet potatoes, turnips, and more—grow on thin strips of land, covered in black plastic to keep down weeds (because he doesn't use chemical weed killers). He rotates his crops aggressively, so that each season a different plant is growing in a different parcel. This rotation is a healthy alternative to spraying insecticides because pests tend to specialize on one type of plant. He explains that if you plant the same crop in the same place every year, the pests that specialize on that plant become stronger and would devastate the crops if they weren't doused in chemicals.

He stoops down and examines the leaves of a red clover, which is growing in a field that recently had turnips. The advantage to growing clover, he explained, is that it and other legumes suck nitrogen gas out of the air and pipe it down into their roots. At the end of the season, Grohsgal will plow the clover under. Then all the nitrogen that the clover collected will serve as nitrogen fertilizer for the next season's crop here. Planting clover and then plowing it back into the soil is a natural alternative to using chemical nitrogen fertilizer, which is a major pollution problem for the Chesapeake Bay.

"So what do we do that's really abnormal?" asked Grohsgal. "We use less than one fifth of the amount of nutrients that farmers are permitted to use under state law to fertilize our crops. And what does that do? I don't have to worry that in a heavy rainfall I'm polluting the Chesapeake Bay. I don't even have to think about it."

Like Fry, Grohsgal raises animals—but in a different way. He guides me to a rickety wooden former tobacco barn, where 400 chickens wander into a yard with tall grass. Grohsgal explains that his free-range chickens lay eggs that taste better than supermarket factory farm eggs because his chickens have a more diverse and richer diet, including bugs and plants instead of just corn meal and antibiotics. His chicken yard is fenced in but relatively exposed. For this reason, foxes and other animals sometimes sneak in, kill the chickens, and steal the eggs. This higher mortality rate is one reason his eggs are twice the price of supermarket eggs. On most factory-style farms, chickens lay their eggs in steel buildings that no predators can enter.

Grohsgal is also different from many conventional farmers today in that he doesn't accept any government money. Most grain farms receive federal price supports or subsidized crop insurance for corn and soybean production, which—in the end—mostly help big companies like Pepsi and General Mills.

"To a great extent, the soybean, feed corn, and winter wheat economy in the U.S. is still dependent to this day on federal Farm Bill money," Grohsgal said of the national system of government crop subsidies.

The Farm Bill is really about entitlement for grain farmers. And it's a bad situation, because it has made grain farmers essentially addicted to these handouts. At the same time, it powers huge companies—like Cargill and Kellogg's and the people who bring you high-fructose corn syrup. They are basically getting cheap, subsidized grain at the taxpayer's expense and they are able to turn that into soda or into all of the different sweeteners that now pop up in potato chips and other junk foods. The people who make most of the money are the food processors, and they don't want the Farm Bill to end.

Caught up in this system are most consumers, whose health suffers because of obesity and diabetes caused in part by this river of cheap high-fructose corn syrup that companies sneak into just about every drink and processed food sold in supermarkets.

Grohsgal's farm is the enemy of obesity—both because of the vegetables and fruits he grows and, in a more immediate sense, because he and his workers must sweat and toil to weed the fields by hand. They use their muscles instead of spraying weeds with herbicides.

Farther down the gravel road, five workers were on their knees on black plastic sheets covering a tomato field. As one of the workers planted, he stood up, bare-chested and sweating, and wiped his forehead. Grohsgal shook his head, disapproving. "If they are standing up, they can't be working," he said.

Grohsgal may look like a hippie, but he toils like a marine. Hard physical labor, he explains, is another major difference between organic farms and conventional farms. He employs 10 local residents to weed and plant by hand, paying them $9 to $11 per hour. If this were a modern corn farm, he would use machines and chemicals to plant the seeds and keep the weeds down, and so he wouldn't have to employ anyone. In this way, his organic farm is more like a nineteenth-century farm, which demanded much more back-straining human labor, instead of chemical and machine labor.

"Paying people to weed by hand is expensive," Grohsgal said. "And that's the real tradeoff in organic, is that you are paying people to do what a chemical could do."

Unlike conventional corn farmers or milk producers, Grohsgal doesn't sell to brokers or onto a national market. So he's not at the mercy of price fluctuations that are beyond his control. Instead, he created his own network of 500 subscribers who have agreed to buy his organic vegetables, fruits, and flowers. It's a system called "community-supported agriculture" (CSA). Grohsgal and his employees truck loads of produce to several different locations around the state for his subscribers to pick up. "I love the CSA model," he said. "My wife and I did not start this business so that we could be selling through brokers who would take our food, mix it with a lot of other food, and then pass it on to the consumer—and keep most of the money. This farm works better, and it has been profitable." He said he grosses about $300,000 per year, but that the majority of this revenue goes to the payroll of his workers.

We stroll past fields of gladiolas, which are buzzing with fat bees—a sign that no neonicotinoids or other bee-killing insecticides are being used on his farm.

"Farmers who use chemicals heavily often have to cope with huge amounts of chemical overloads in their own systems," Grohsgal said. "So many a conventional farmer suffers from diseases later in life."

He leads me across a field and shows me that in the middle of his farm he planted rows of black locust and loblolly pine trees flanking a pair of streams. These strips of forest act as green filters to catch any runoff that might otherwise pollute the creeks and bay. He guides me through the line of trees, past a fallen giant with mushrooms popping from its exposed roots. Finally, we see a brook that meanders its way through the farm's woodlands.

"The cool thing about this stream?" Grohsgal asks. "If you come down here in the middle of a hurricane, you can still see through the water. It's not cloudy with sediment and mud and soil that's been washed off of the fields. It runs clear and clean, even in a storm."

To try to encourage more forested buffers like this, the Obama administration in 2010 set a goal of using financial incentives to encourage farmers across the Chesapeake Bay region to plant a collective total of 900 miles of trees along streams per year.[25] But farmers figured out they could make more money—and face fewer hassles—if they just kept their waterfront land in corn and soybeans. So they have been volunteering for only about 100–150 miles of these forested buffers per year, meaning that the bay cleanup effort is far behind.[26]

"We are trying to bring the numbers up, because forested buffers are the best land cover to have next to a stream," said Sally Claggett, Forest Service liaison with the Chesapeake Bay Program.[27] "The trees provide the last line of defense as water is coming off the land and entering a waterway, the last opportunity for that water to be filtered."

Some farmers say they don't like to sign the contracts required to receive government subsidies for forested buffers because the contracts put restrictions on the farms. These restrictions include requiring the property owners not to build on or farm the strips of land beside streams for 15 years. Farmers bristle at the idea of taking a portion of their cropland out of production and hate the idea of losing money because a part of their land can't be sold and developed.

Because of the slowness of the Chesapeake states to push farms to improve their environmentally friendly practices, the bay is not on track to meet the EPA's pollution reduction targets for 2025 (the so-called bay pollution "diet" or Total Maximum Daily Load), said Tom Simpson, a veteran bay watershed agricultural expert and former senior scientist at Water Stewardship Inc.[28]

"The level of effort will have to increase immensely in all sectors, but particularly in agriculture," Simpson said. "The amount of implementation is going to have to increase dramatically. And probably there is going to need to be some systemic changes and retirement of a small percentage of farmland in critical areas if we are going to achieve our goals—not just by 2025 but forever. We are far behind and not gaining."

The farms that are getting ahead are often those like Brett Grohsgal's that are going back to the future. And the ones that are struggling most are those—like Chuck Fry's—that are stuck in a modern, industrial system that burdens both the farmers and the bay.

CLIMATE CHANGE

At War with a Changing Climate

A T THE FAR SOUTHERN END of the Chesapeake Bay, Jonathan White, a retired rear admiral in the US Navy, was in a van giving a tour of the world's largest naval base, Naval Station Norfolk. While some people dismiss climate change as a hoax, White, an oceanographer who spent 32 years in the Navy, including as director of the Navy's Task Force on Climate Change, said that the military is well aware of the serious and well-documented threat posed by sea-level rise caused by warming temperatures.

"If we just did nothing with this base, it's easy to see that in a worst-case scenario of a couple of meters of sea level rise by the end of the century, much of this base would not be usable," said White, as he passed a dock occupied by the nuclear-powered aircraft carrier USS *George Washington*.[1] "You wouldn't even be able to get to the ships to pull them in to the piers here. Even a *moderate estimate* of a meter of sea level rise is going to create a lot of issues to continue to use this as a Navy base."

Abandoning the century-old military base—with its 14 piers, 75 ships, 11 aircraft hangers, and constellation of surrounding facilities—to rising waters would also cause a few issues (to put it mildly) for the 1.7 million people who live around Hampton Roads, because almost half of the local economy is linked to the defense industry.[2] Tens of thousands of local jobs could be lost.[3] But that's only the tip of the iceberg. Hampton Roads is experiencing the highest rate of relative sea-level rise on the East Coast, and as many as 176,000 of its residents live in low-lying areas that would be flooded by a rise in water levels of 3 feet or more over the next century.[4] Many of the Chesapeake Bay's waterfront areas—including downtown Annapolis; Baltimore's Fells Point and Inner Harbor neighborhoods;

and Tangier, Smith, and Hooper's Islands—are vulnerable to inundation from higher storm surges caused by warmer temperatures expanding the volume of the oceans.

Climate change is also triggering an array of other chain reactions—some complex and profound—in the Chesapeake Bay region. Average water temperatures in the estuary have already risen almost 3 degrees Fahrenheit since 1940 and could jump as much as another 9 degrees by 2100, according to a report by the Maryland Commission on Climate Change.[5] Average summer air temperatures in the region are also expected to surge by as much as 9 degrees this century, with as many as 24 days each summer exceeding 100 degrees by 2100, triggering more deaths from excessive heat among the elderly and poor.[6] From these climbing temperatures may spin a curious combination of more summer droughts and more intense rain- and windstorms.[7] More pollution flushed off the land by these big rainstorms is likely to worsen water quality and expand low-oxygen "dead zones" in the bay. Low-lying islands and wetlands are already being swallowed by rising waters. The damage is especially acute at the Blackwater Wildlife Refuge on Maryland's Eastern Shore, where more than 5,000 acres of wetlands have disappeared since the 1930s.[8] Warmer temperatures are bringing more southern species into the bay region, such as the brown pelican, a bird native to Florida and South Carolina that has been nesting in the Chesapeake only since the 1980s. Biologists have found that osprey, a fishing bird that historically has nested in the Chesapeake and migrated every winter to the Caribbean and South America, are beginning to overwinter in the bay for the first time because of the warmer conditions. The unnaturally warm temperatures early in the spring are also shifting the migratory calendar of striped bass. A freakish boom in red drum—a fish common in the Gulf of Mexico—in the Chesapeake in 2012 was blamed on a decline in blue crab populations in the bay, as red drum eat juvenile crabs. As these southerners flood in, other native critters that prefer a more northern climate are being driven out, including soft clams, eelgrass, and Baltimore orioles. The exit of this eelgrass, in particular, could hurt blue crab populations, because juvenile crabs rely on eelgrass beds in the southern bay to hide from predators. More carbon dioxide in the atmosphere is also spawning a chemical reaction that is creating carbonic acid, which is making the waters more acidic. The result is the thinning of shells of oysters and clams, which could potentially have an impact on the Chesapeake Bay's rapidly growing shellfish farming industry.

"There is a big problem with shellfish, both from the temperature and from acidification of the water," said Skip Stiles, executive director of Wetlands Watch, a nonprofit organization based in Norfolk. "Virginia is the number one hardshell clam aquaculture state in the country right now. There are people making $50,000 per bottom acre on the Eastern Shore. So as climate change begins to affect the clam industry—yeah, there is going to be some economic impact."

For most people who live near the water, the most immediate danger from the Chesapeake's changing climate is more flooding during storms. In few places is the risk of deluge more serious than in southeastern Virginia, where the Chesapeake's naturally sinking lands are subsiding faster than elsewhere. An 8-inch rise in global ocean levels over the past century has been compounded in the Hampton Roads area by a roughly 1-foot drop in the ground level.[9] This settling of the land is being caused by the natural shifting of geological formations, as well as by the extensive pumping of groundwater from drinking water aquifers beneath Hampton Roads.[10]

"So you have the global sea level rise—but then on top of that here you also have the subsidence of the land," said Admiral White. "It's a little bit like when you're standing on a beach, and the water comes in and your feet sink into the sand. That's what makes the Hampton Roads area, as well as some areas on the Gulf Coast—like New Orleans and Galveston and places like that—extremely vulnerable to sea level rise, because they are sinking at the same time that the water is rising."

After driving around the base, we toured the neighborhoods of Hampton Roads on an educational trip organized by the nonprofit World Resources Institute. Our van blasted through a half foot of standing water that swamped the roadways at several locations. It was not high tide, and it had only rained a bit the night before, but whole streets, lawns, parking lots, and communities of Norfolk were flooded—again, as they are several times a season.

"It used to be just once or twice a year it would flood here," White said, gazing down the street at an entire city block submerged in water. "Now it's probably flooding 10 to 20 times a year, something like that. In the future, we could be having hundreds of floods like this. That really can have catastrophic impacts on your military readiness—if you don't do anything about it."

Also on the tour was Joseph Bouchard, the former commanding officer of Naval Station Norfolk, now retired. Bouchard recalled that, during his command of the base in the early 2000s, the rising sea levels frequently caused electrical blackouts on the piers, which made it harder for the sailors to load their ships and prepare themselves for combat missions. "As sea level rose, the piers were inundated more and more often. High voltage electrical lines and seawater don't mix very well," Bouchard said. "So we were having increasing outages, and that's a real problem—because a Navy ship in port is not just sitting there, idle. They are always hard at work, doing maintenance and training."

We climbed out of the van at one intersection in Hampton Roads where water blocked all traffic. A construction crew was trying to lift a house up on top of 6 feet of cinderblocks to protect it from future inundation. But the workers were having a muddy slog of it, struggling to operate their backhoe and truck in a 6-inch deep swamp that engulfed the yard and road.

"Most people in the United States have never seen this kind of flooding on their streets or in their front yards. But this is just an everyday occurrence here," said Christina DeConcini, director of government affairs at the World Resources Institute. "We didn't have a hurricane or a Nor'easter or some incredible weather event that we are looking at here. This is just an example of the rising seas."

A few blocks away, Henry Braithwaite, a Norfolk construction contractor, shook his head at the standing water that was again occupying his street. "I can't even get into work today because there's no way I can back my car up to get out of my driveway," Braithwaite sighed, as he stood beside his home in his bare feet with a cup of coffee in hand. "It's come to the point where this is a regular nuisance. I'm not sure where all this water is coming from, because it didn't even rain that much last night."

Down at a waterfront neighborhood called the Hague, the canal had breached its banks and cars slowed to splash through a pond that blocked the entrance to the Unitarian Church of Norfolk. A large "for sale" sign stood on the church's lawn, an indication that the worshippers had lost faith in waging war against the oceans and had decided instead to flee.

A neighbor, Rachel Barr, was walking her dog beside the new pond and stopped to express astonishment at how much the city had changed.

"What's amazing is that, so far we've only seen about a foot and a half of sea-level rise" over the past century, Barr said. "We are expecting at least another *three feet* rise over the next century. What's going to happen then?"

Not far away, at a waterfront neighborhood along the Elizabeth River, the city of Norfolk is planning to build an earthen berm, artificial wetlands, and other projects to try to slow the water's rise, using $120 million in federal funds. But of course, in the long run, a pile of dirt won't do a heap of good to stop an ocean. The real answer is burying the fossil fuels that created the problem in the first place.

ADVOCACY AND POLLUTION TRADING

How "Save the Bay" Became "Trade the Bay"

THE IDEA OF "saving the bay" was popularized by the Chesapeake Bay Foundation (CBF), a nonprofit established in 1967 that has grown over the past half century to become the largest regional environmental organization in the United States.[1] It is hard to overstate the influence of CBF on the bay restoration effort and public consciousness, with the foundation's "Save the Bay" bumper stickers on cars and refrigerators everywhere. In fact, when people in this region—especially school children, but also the elderly—think about helping the environment, often the very first thing that pops into their minds is to donate to or raise money for CBF. The animating principle of the Annapolis-based organization, where I worked for almost six years from 2008 to 2014 as senior writer and investigative reporter, is that we can "Save the Bay" if we all work collaboratively to make incremental progress, with Republicans and Democrats, conservative farmers and liberal urbanites, scientists and watermen all agreeing to cooperate. But not all donors know that the voluntary, nonregulatory approach favored by CBF—especially for farms, the biggest source of pollution in the bay—is very different from the one taken by most other environmental groups. Decades of evidence have shown that top-down regulation of polluters has proven to be more effective than voluntary methods. CBF, however, has always preferred a nonconfrontational, Republican-friendly posture in part because it was founded by a Republican attorney and yachting enthusiast from Baltimore, Arthur Sherwood, after a pivotal meeting with a Republican congressman from the Eastern Shore, Rogers C. B. Morton, who was later chosen by President Richard Nixon to chair the Republican National Committee.[2]

CBF's bipartisan approach is appealing to many people, especially in an era of hyperpartisan name-calling and gridlock in Washington, DC. And the idea made sense back in the 1970s and 1980s, when Republicans were just as likely as Democrats to support environmental regulations. All honest Democrats must acknowledge that it was a Republican president, Richard Nixon, who created the EPA and signed the landmark Clean Water and Clean Air Acts, which have been the engines of most environmental progress in America over the past four decades. It was another Republican president, George H. W. Bush, who signed the 1990 amendments to the Clean Air Act which further reduced air pollution, and Republican Maryland governor Robert Ehrlich who imposed the "flush tax" in 2004, which worked to upgrade most of the state's largest sewage treatment plants. Going way back in time, it was Republican presidents Abraham Lincoln and Theodore Roosevelt who invented and expanded America's National Park System.

However, in more recent years, bipartisanship in support of environmental goals has become an endangered species driven toward extinction. Most Republican elected officials in favor of environmental programs vanished as soon as it became clear that climate change could only be addressed through national regulation of the coal and oil and gas industries, which donate heavily to their campaigns. Today's Republican Party is anti-regulatory and anti-government with a fierce, ideological purity that makes Nixon and George H. W. Bush look like liberals in contrast. Anti-environmentalism today has become a litmus test of party purity for Republican elected officials, like being anti-abortion or anti–gun control. Moderate and pro-environment Republicans (like Representative Wayne Gilchrest of Maryland) have been voted out of office.[3] Republican president Donald Trump, during his campaign, falsely called climate change a "hoax" invented by the Chinese and advocated gutting the EPA and eliminating 70 percent of regulations.[4] Once he was elected, Trump proposed draconian funding cuts to the Chesapeake Bay Program.[5] The Republican-led US House of Representatives voted in 2016 to strip the EPA of all power to enforce pollution limits for the Chesapeake Bay by penalizing states that did not meet their cleanup goals.[6] In this radically changed political climate, CBF continues to dance its old bipartisan dance—although its partner left the floor a long time ago.

CBF's nonconfrontational strategy has been warmly welcomed by those in power (including corporations), in part because the foundation offers itself as a moderate alternative to liberal environmental groups that shout for change. CBF's positioning has worked phenomenally well for fund-raising, with conservative waterfront estate owners, companies like Walmart and Pepsi, government agencies, and many others contributing more than $20 million a year to the organization.[7] The foundation's financial statements burst with assets that totaled an astounding $107 million in 2016 (including $57 million in stocks, bonds, and other investments).[8] CBF has expanded over the years to employ about 170 people in a growing network of six offices and five education centers spread across the region. It is fair to say that CBF has been spectacularly successful— but that the organization itself, with its $17 million waterfront headquarters in Annapolis, has prospered far more than the bay outside its windows, whose health has hovered just above bankruptcy.[9] CBF's offices boast a beautiful beach, but its employees rarely swim there, in part because of well-founded fears of high bacteria levels in the murky olive-green waters.

The bay desperately needs a well-funded army to fight for its life, but instead it has a flagship organization that takes a much softer approach and prefers quiet backroom deals to public confrontation of those in power. Here are a few examples. When Maryland governor Martin O'Malley from 2011 to 2014 repeatedly delayed regulations to stop the poultry industry's dumping of excess poultry manure on Eastern Shore farm fields, CBF declined to publicly criticize O'Malley because he was an important political ally. O'Malley wanted to appear farm-friendly to win votes in the Iowa caucuses during his failed quest to win the Democratic presidential primary. CBF also shied away from making a public outcry when O'Malley diverted millions of dollars in state funds that were supposed to be dedicated for land preservation and bay cleanup.[10] In Pennsylvania, when a succession of governors effectively gave the finger to Maryland and other downstream neighbors by failing to enforce Pennsylvania's Clean Streams Law and manure management requirements for farmers, CBF praised the governors or remained silent about the lack of penalties for the widespread violations. Even as oyster populations in the bay plummeted to *one-third of 1 percent* of historic levels, CBF gave up on the idea of advocating for a moratorium on harvesting oysters because a ban might prove controversial among watermen. Instead, CBF worked with a seafood industry–friendly group called the "Oyster Recovery Partnership" and supported incremental changes instead of the ban on oyster dredging for which the bay is crying out. The Oyster Recovery Partnership, it should be noted, has as its chairman Jim Perdue, chairman of Perdue Farms Inc. (a notorious polluter of the bay), and as its vice chairman Russell Dize, who is also vice chairman of the Maryland Watermen's Association (which lobbies against any restrictions on harvesting oysters).[11]

CBF has made some strong moves to help the bay in recent years. For example, the organization deserves a tremendous amount of credit for successfully suing the EPA in 2009 to accelerate the federal agency's imposition of pollution limits and reduction targets for the bay states. As a result, the Chesapeake Bay watershed is the only one in the United States that has a multistate, EPA-monitored pollution reduction "diet"—the so-called Chesapeake Bay Total Maximum Daily Load or TMDL. Among other noteworthy accomplishments, CBF also played important roles in helping to stop a $1 billion golf resort planned near the entrance to the Blackwater Wildlife Refuge on Maryland's Eastern Shore (in 2007) and halting a reservoir project in southeastern Virginia, the King William Reservoir, which would have destroyed 430 acres of wetlands (in 2009).[12]

CBF's most questionable partnership in recent history has been its political alliance with the Maryland Farm Bureau. A "Save the Farm, Save the Bay" coalition was announced at a September 20, 2005, press conference on a farm in Prince George's County. "This is history in the making— the Maryland Farm Bureau and the Chesapeake Bay Foundation standing together," CBF president Will Baker pronounced, standing beside Farm Bureau president Buddy Hance.[13] The alliance came despite the fact that agriculture is the single largest source of pollution in the bay and the farm lobby is the most strident opponent of clean water regulations. It would be somewhat like a West Virginia environmental group forming an alliance with the coal mining industry, with the understanding that the green group would receive money to plant trees on former mining sites. There is nothing wrong with planting trees, of course. But such a "green" organization might not feel free to criticize its partner when it ripped the tops off of mountains.

What did CBF's partnership with the farm lobby mean? CBF successfully lobbied Congress for millions in federal Farm Bill funds to encourage Chesapeake region farmers to install pollution control projects on their farms, and CBF itself received at least $19 million in federal funds over a decade to work with farmers.[14] CBF has about 10 employees working directly with farmers, trying to convince them to join voluntary programs to receive government money to fence their cattle out of streams, plant trees along waterways, and take other voluntary actions to reduce runoff. On the positive side, because of these outreach efforts, hundreds of farmers have taken steps to improve local water quality.[15] On the negative side, tens of thousands of other farmers do not have to worry about the region's largest environmental organization advocating for regulations to require these same pollution control projects for everyone. CBF issues reports on all kinds of bay-related issues—but avoids aggressive reporting on agricultural pollution because it does not want to anger the farmers who are their clients and the farm-allied elected officials who approve the funds for which it lobbies.[16]

CBF often—although not always—diverges from other environmental groups on agricultural policy. For example, in May 2008, when Governor O'Malley's administration weakened proposed pollution control regulations for factory-sized poultry houses, all of the other environmental organizations in Maryland condemned the move, but CBF praised O'Malley.[17] In 2013, CBF successfully championed an "agricultural certainty" bill that grants Maryland farmers a 10-year exemption from any new environmental regulations if they meet basic standards for controlling runoff.[18] The law had been supported by farmers seeking fewer pollution control rules, although the decade-long exemption could hurt the bay's water quality. In a move by CBF which was praised by the National Chicken Council, National Turkey Federation, and US Poultry & Egg Association but condemned by the Sierra Club, Environment Maryland, Waterkeepers Chesapeake, and other environmental groups, CBF in 2013 struck a deal with the EPA in which the federal agency backed away from an earlier commitment, made in response to a CBF lawsuit, to impose new pollution control regulations on large factory farms.[19] In October 2016, CBF praised an announcement by the US Department of Agriculture, the EPA, and the state of Pennsylvania that the government agencies and state would direct another $28.7 million in taxpayer funds to farmers in Pennsylvania for runoff control projects. But CBF's press release did not mention the obvious law enforcement problem with these farms.[20] Almost half of Pennsylvania's livestock farms in the bay watershed do not have or follow manure management plans that were required by state law more than three decades ago.[21] The state should be enforcing this law and fining farmers who allow their manure to pollute public waterways, not simply handing yet more taxpayer money to the farmers. (Many of the rural conservatives who receive government money, ironically enough, strongly oppose using government money to help needy urban residents and demand strict law enforcement in cities—just not on farms.)

In the myriad of issues that swirl around the bay, CBF is often not on the farm lobby's side. For example, in 2011, the American Farm Bureau, developers, and others unsuccessfully sued the EPA over the pollution limits the agency imposed on the bay region states (although, oddly enough, this federal pollution "diet" imposed no regulations on farmers). CBF filed a motion in support of the EPA, not the Farm Bureau. In 2016, the foundation joined environmentalists in supporting a bill that would have held big poultry companies responsible for the cost of properly disposing of excess manure. The legislation never made it out of committee.

Looking to the future of the bay, perhaps the most important agricultural policy question revolves around pollution trading. Trading is a strategy that is supported by both CBF and the Maryland Farm Bureau in part because it directs government money to farmers for pollution control projects and is seen as an alternative to traditional, top-down regulation of farms. In 2012, the US Department of Agriculture awarded CBF a $700,880 grant to help design and encourage water quality trading as a uniform policy across the bay region, including by "promoting the use of a consistent tool for estimating nutrient reduction credits from agricultural operations," according to grant documents.[22]

So what is this policy that CBF is promoting? Pollution trading is a scheme first conceived by Republicans in the 1980s, when the Reagan administration held up free markets and deregulation as medicines for all that ails America. (The same philosophy gave birth to charter schools, privatized prisons, privatized municipal water systems, and other market-based alternatives to government services.) Back then, sulfur dioxide pollution from coal-fired power plants was causing an increasing amount of acid rain that was damaging forests and streams. But President Reagan was not about to impose traditional regulations, because he defined government as the problem, not the solution. So Vice President George H. W. Bush's legal counsel, C. Boyden Gray, thought up a more politically palatable idea that carried the glow of Wall Street: pollution trading.[23] Power plants and other industries could buy and sell the right to pollute. But excessive emissions would be discouraged because they would come with a price tag. Each power plant would be given a cap—or maximum amount—of sulfur dioxide it would be allowed to spew into the air. If plants wanted

to release more than that, they would have to purchase credits from someone else on the free market. The concept was incorporated into the 1990
amendments to the federal Clean Air Act and signed by the first President
Bush. This pioneering "cap-and-trade" system worked to drive down sulfur dioxide emissions and reduce acid rain, and was a great success. However, the same reductions could have been achieved more quickly by simply
requiring that all power plants install pollution control devices, including
scrubbers. That was the top-down regulatory approach taken, for example, by Germany in the 1980s to achieve the same results.

Since the 1990s, cap-and-trade systems have been proposed as a magic
bullet for a variety of other pollutants, including carbon dioxide in the
atmosphere and nitrogen in the wastewater from sewage plants. The
evidence so far seems to suggest that cap-and-trade systems can work, if
both of the trading partners have smokestacks or waste pipes that are
carefully monitored, so that regulators can make sure that an increase in
pollution from one source is more than compensated by a decrease from
the other. There is no evidence, however, that trading works when the
pollution cannot be monitored—for example, runoff pollution from farms
or suburban streets. Despite this problem, however, over time the idea
of pollution trades between unmonitored sources has been embraced by
state governments in Pennsylvania, Maryland, and elsewhere. The once-
Republican concept of applying a Wall Street mentality to nature was
embraced by some Democratic officials as the party shifted rightward in
the 1990s and 2000s. In the same way, the deregulation of banks and Wall
Street, originally a GOP cause, was eventually advanced by the Democratic Clinton administration.

The main argument in favor of pollution trading made by CBF and
others is that it can be cost-effective: the bay could, in theory, get more
cleanup "bang" for the buck through strategic pollution trades.[24] Reducing
a pound of nitrogen pollution in the stormwater pouring off of an urban
area, for example, can be tremendously expensive, because it requires the
ripping up of streets and parking lots to build filtration systems into the
streetscape. Eliminating that same pound of nitrogen pollution from
farm runoff can be much cheaper, simply requiring a farmer to spread less
manure on his land, or to back up from the stream his planting of corn
and spreading of fertilizer. If a city could use its money not to rip up
streets and parking lots for stormwater control systems but instead to buy
nitrogen reduction "credits" from farmers, the end result could be more

pollution kept out of the bay. In 2013, CBF senior scientist Beth McGee testified to Congress in favor of trading because of this cost efficiency.[25] And in 2016, the administration of Maryland governor Larry Hogan pushed to make pollution trading a reality in the state. "The beauty and promise of water quality trading is . . . more cost-effective ways to reduce excess nitrogen, phosphorus and sediment—real threats to the Chesapeake Bay," said Ben Grumbles, Hogan's secretary of the environment.[26]

The problem is that, under this kind of pollution trading, waterways in urban areas like Baltimore could get no cleaner—while the tax dollars of lower-income urban residents would be shipped off to improve the waterways and quality of life of often wealthier rural residents (for example, by planting trees along streams in places like Harford County). And that is assuming that the trading even works. In the case of agricultural pollution, effective monitoring—and therefore confirmation of success— is almost impossible because the runoff flows from millions of different places, including slowly through groundwater in the crevices between soil particles. To make matters worse, farmers demand that the records of their pollution control efforts be kept secret. So there would be no way that scientists, environmental agencies, or other independent researchers could check whether pollution reduction efforts purchased as part of a trade are real or exaggerated. The backbone of the Chesapeake cleanup effort—the enforceability of the federal Clean Water Act—could break, because the trail of evidence would disappear.

Without monitoring and transparency, there can be no accountability and therefore no legitimacy for pollution trading. CBF argues that it favors accountability and transparency measures for pollution trading,[27] but it also contradicts this claim by endorsing trades between farms and other "nonpoint" sources of pollution, in which it is inherently impossible or very difficult to accurately measure reductions in pollution, because of the diffuse, delayed, and unmonitored nature of stormwater runoff. Concerns about the legitimacy of pollution trades are not idle speculation. Back in 2007, when the EPA used a market-based credit trading system to create a financial incentive for producing biodiesel, a vegetable-based alternative fuel, the result was that con artists in Maryland and elsewhere took advantage of the flexibility in the system to sell millions of dollars in fraudulent credits for imaginary air quality improvements.[28] A European carbon dioxide pollution trading scheme dissolved into accusations of fraud in 2009.[29] Rigorous monitoring and oversight are needed to make trading work, but Pennsylvania was criticized by the EPA in 2012 for launching a water pollution trading program that lacked even the most minimum controls and safeguards.[30] In a planning document that Pennsylvania officials submitted to the EPA that year, state officials were candid in admitting that they hoped that their pollution trading system would be a mechanism by which sewage plants could *avoid* modernizing, by simply paying to pollute. "Buying the credits may help the plants avoid upgrades entirely, or allow them to do less expensive upgrades," Governor Tom Corbett, a Republican, wrote in a Pennsylvania bay cleanup plan in 2012.[31]

We should be making it harder, not easier, for states to avoid upgrading their sewage treatment plants—the most proven method of cleaning the water. Instead of absurdities like pollution swaps that cannot be monitored, the bay needs strong—and strongly enforced—rules that require sewage plants, farmers, and everybody else to reduce their pollution. And the Chesapeake Bay deserves a flagship advocacy organization with the backbone and political courage to stand up and oppose a scheme—such as pollution trading—that undermines accountability for polluters.

ACCOUNTABILITY

The Bay Numbers Game

MOST OF THE three-decade-plus history of the bay restoration effort has been like a football game, in which—after a few quarters—the teams get bogged down around the 50-yard line with no score. So the officials move the goalposts closer. Then they quietly move the posts again.

When it's a struggle to clean up the bay, sometimes it's easier just to clean up the numbers.

Here's an example of the bay's creeping goalposts: Back in 1987, the EPA signed a bay cleanup agreement with Governor William Donald Schaefer of Maryland, Mayor Marion Barry of Washington, DC, Governor Gerald Baliles of Virginia, and Governor Robert Casey of Pennsylvania in which all of the parties pledged to reduce their nitrogen pollution into the estuary by 40 percent by the year 2000.[1] But as Tom Horton pointed out in his book *Turning the Tide*, shortly after the goal was set, "40 percent" was redefined to mean "40 percent of *controllable* sources," which arbitrarily excluded half of all nitrogen sources—including from air pollution (although this is controllable)—and discharges from New York, West Virginia, and Delaware, which were not signatories to the agreement.[2] With this subtle tweaking of the language, the goalpost for 2000 suddenly moved much closer—essentially requiring a 21 percent reduction in nitrogen pollution instead of a 40 percent cut. Unfortunately, this more modest goal was still not achieved by 2000, so the bay states signed a new agreement with a new set of targets—which were changed again in 2010 after another decade in the dust near the 50-yard line. "It is unlikely, even if the bay continues for much longer in an unrestored state, that political leaders and decision makers will admit failure," Horton wrote in 2003. "What is likely is that they, and we who live around the bay, will redefine success and change the concept of 'restored.'"[3]

All these years later, Horton has proved himself the bay's oracle. None of the bay's major players have admitted failure, despite billions spent with only slight progress. But they have redefined success in such ingenious terms that in a few years they may be able to declare victory and spike the ball even if the water is as murky as ever.

The background is this: Following the 1987 bay agreement, the bay region state governments, led by Maryland governor Parris Glendening, signed another "historic" compact, the Chesapeake 2000 agreement.[4] Like its predecessors, the Chesapeake 2000 model was strong on rhetoric— declaring the bay "a resource of extraordinary productivity, worthy of the highest levels of protection and restoration"—but weak on muscle. It was a purely voluntary statement of lofty ambitions, with no commitments by the states to enact any of the regulations that would be necessary to achieve the targets, and no mechanism by which the EPA could hold the states accountable. In its ambition and bureaucratic puffery, however, the Chesapeake 2000 agreement was monumental. Beginning with the pronouncement "*We Commit To*" in the florid script of an eighteenth-century treaty between King Louis XVI of France and George III of England, the agreement set forth 87 bulleted goals. The majority of these had to do more with bureaucratic process—the writing of more plans, goals, and assessments—than the execution of changes in the real world that would actually make the bay cleaner. Six of the goals, however, set numeric targets and dates for concrete improvements that would (if accomplished) achieve significant progress in cleaning up the water.

The states met some targets—for planting trees and preserving land, for example—but failed to meet four of these six numeric commitments by the deadline of 2010. So they worked with the EPA to create yet another agreement with new goals. But before I get to that, I am going to run through some of the specifics of the Chesapeake 2000 agreement, because they are illustrative of the troubles that roiled the bay during a decade (2000–2010) in which there should have been forward momentum, but during which there was mostly backsliding.

Oysters. The very first goal established by the Chesapeake 2000 agreement was for a 10-fold increase in oysters in the bay by 2010.[5] This turned out to be a huge failure. Taxpayers invested millions of dollars to plant baby oysters over this decade. But unfortunately, the bay agreement did not commit the states to any regulations that would protect the oysters from oystermen. As a result, as noted earlier in this book, the number of oysters in the bay actually plummeted by about 70 percent from 2000 to 2010 instead of multiplying 10-fold, in part because Maryland and Virginia allowed watermen to continue harvesting the oysters in a destructive and wasteful "put and take" fishery.[6] The result was that the population of the shellfish plummeted to a near-extinction level of *one-third of 1 percent* of historic levels by 2009, with disease also contributing to the decline, but playing a smaller role than chronic overfishing.[7] That year, Governor Martin O'Malley's administration changed directions on oyster policy and protected 24 percent of the bay's remaining oyster reefs with sanctuaries, which allowed oyster populations to start slowly increasing again.[8]

Nitrogen and phosphorus pollution. In the Chesapeake 2000 agreement, the bay region states pledged, by 2010, to "achieve and maintain the 40 percent nutrient reduction goal agreed to in 1987."[9] As discussed earlier in this book, nitrogen and phosphorus come from fertilizers, sewage plants, and other sources and feed algal blooms and "dead zones" in the bay. Largely because of upgrades to sewage treatment plants and declining air pollution, the bay region states achieved roughly half of the nitrogen reduction goals of the 1987 agreement by the original deadline of 2000 (using even the modified, watered-down definition of "controllable sources" of nitrogen).[10] However, between 2000 and 2010, even that limited progress hit rough waters and stalled. By one method of calculation, the amount of nitrogen pouring into the bay actually increased by 18 percent during this decade, rising from 258 million pounds in 2000 to 304 million pounds in 2010, according to EPA Chesapeake Bay Program estimates.[11] The amount of phosphorus pollution nearly doubled, increasing from 11 million pounds in 2000 to 21 million pounds in 2010. However, a note of explanation is needed for these figures: significantly more rain fell in 2010 than in 2000, so that would likely explain much of the increase, as more precipitation flushes more fertilizer and other pollutants off the land and into the bay.

To get a more accurate picture of what happened over this time period, let's compare two years with similar amounts of rainfall, 1990 and 2006. In 1990 an estimated 338 million pounds of nitrogen flowed into the bay, and in 2006, it was 320 million pounds.[12] That would mean a 5 percent reduction in nitrogen pollution, compared to the goal of a 40 percent cut, according to EPA Bay Program estimates.[13] For phosphorus, the picture was worse, with a nearly 12 percent increase in pollutant loads entering the bay, rising from 15.6 million pounds in 1990 to 17.4 million pounds in 2006.[14]

What does this all mean? Despite some progress in reducing nitrogen pollution from sewage plants over this time period, phosphorous levels did not improve much (if at all). And, more broadly, the big picture is that sewage accounts for only a small portion of the pollution in the bay: about 15 percent of the nitrogen and phosphorus. Urban and suburban runoff is just as large of a problem as sewage, and farm runoff is about *three times* more important than either one of these issues, with farms accounting for 42 percent of the nitrogen pollution in the bay and 55 percent of the phosphorus.[15] This runoff pollution is also far more challenging to tackle, in part because the laws that regulate it are weaker (in particular, the federal Clean Water Act has a giant loophole for grain agriculture and only flawed control mechanisms for livestock farms). Animal agriculture produces hundreds of millions of pounds of manure, laden with phosphorus and nitrogen, which gets washed by rain into the bay every year. The states have made far less progress on controlling animal waste than human waste, and runoff from suburban sprawl is also a growing burden. One result is that, for a complex mixture of reasons, water clarity in the bay worsened during the Chesapeake 2000 agreement. The rating for the bay's clarity fell from 41 out of 100 in 1986, to 14 in 2000, to a record low of 6 in 2011, according to University of Maryland data from monitoring.[16] The bay just kept getting murkier.

Underwater grasses. The Chesapeake 2000 agreement called for the states to protect and restore 114,000 acres of submerged aquatic vegetation by the year 2010.[17] The states and the EPA later raised this goal to 185,000 acres.[18] Underwater grasses are among the most important indicators and producers of water quality in the bay because they filter out sediment and release oxygen that is vital for bay life. But the grasses cannot survive in waters that are so murky that they block light needed for photosynthesis. At one time, an estimated 600,000 acres of grasses swayed in a vast underwater forest in the Chesapeake, sheltering baby blue crabs and cleaning and bubbling oxygen into the waters.[19] Over the decades, a growing amount of silt and algae in the bay gradually shrank these grass beds. Then, in 1972, the violence of Hurricane Agnes buried huge amounts of vegetation in one muddy rampage. Fewer than 50,000 acres of grasses were left by the 1980s, but that slowly rebounded to 69,126 acres by 2000.[20] Over the next decade, the extent of grasses swung wildly from year to year—with varying weather conditions, rainstorms, and extreme heat from climate change occasionally wreaking havoc. But overall, during the Chesapeake 2000 agreement, the grasses kept growing to about 79,675 acres by 2010, which was significantly below both the original and revised goals for that date.[21]

Wetlands. In the Chesapeake 2000 agreement, the bay region states made this pledge: "By 2010, achieve a net resource gain by restoring 25,000 acres of tidal and non-tidal wetlands." A decade later the states had achieved a little more than half of this goal, restoring 14,782 acres of wetlands, according to the EPA Bay Program.[22] However, the effort has not been as substantial as it might seem. The states and federal government have been reluctant to stop developers from destroying real wetlands and instead have allowed builders to simply pay for the construction of artificial "restored" wetlands to replace real wetlands. The Maryland Department of the Environment, for example, approved 99.8 percent (5,873 out of 5,883) of the applications for wetlands destruction permits which builders submitted to the agency in 2009, 2010, and 2011, according to state records.[23] Percentages in Virginia and Pennsylvania were similarly high.[24] The problem is that scientists have concluded that these artificial "restored" wetlands are almost never as biologically productive as the real thing.[25] A study by the Maryland Department of the Environment found that only half of wetlands restoration projects in the state were successful.[26]

Forested buffers. To promote the planting of trees along streams on farms to act as natural water pollution filters, the Chesapeake 2000 agreement included a commitment by the bay region states to create 2,010 miles of forested buffers.[27] The agreement then called on the states to establish a second, more ambitious target for expanded mileage. The states quickly exceeded this first goal, with 6,582 miles of forested buffers planted in the bay watershed between 2001 and 2010, or an average of 658 streamside miles per year.[28] The Bay Program then boosted the goal to 900 miles of new forested buffers each year. Many of the states made the expansion of forested buffers a key part of their bay cleanup plans. But then farmer participation in the buffer program dropped off dramatically, tumbling from 359 new miles in 2010 to 64 in 2015.[29] Some farmers complained that—with high corn prices, boosted by the federal ethanol mandate—the federal government was not paying them enough money to take their streamside land out of crop production for trees.

Land preservation. In the Chesapeake 2000 agreement, the bay region states pledged to "permanently preserve from development 20 percent of the land area in the watershed by 2010."[30] A decade after making this commitment, Maryland, Virginia, Pennsylvania, and the District of Columbia proudly reported that they had more than surpassed their goal, protecting 7.3 million acres, exceeding the 6.8 million acres that represented 20 percent.[31] However, almost 99 percent of this goal (6.7 million acres out of the target of 6.8 million acres) *had already been preserved before the agreement was signed in 2000*, making the original target less impressive than it seemed.[32] More importantly, far more land could—and by law *should*—have been preserved in Maryland during this decade if Governors Robert Ehrlich and Martin O'Malley had not diverted $466 million from what was supposed to have been a dedicated fund called "Program Open Space" and instead used the money for the general operations of government.[33] Lawmakers created Program Open Space in 1969 to impose a transfer tax of 0.5 percent on all real estate transactions, with all of this revenue directed to protecting forests and fields, creating new parks, and building playgrounds. Since the 1980s, a succession of Maryland governors (with the exception of Parris Glendening, who honored the land preservation program) have diverted more than $1 billion from Program Open Space that should have been used to preserve land and maintain natural green filters around the Chesapeake Bay.[34]

With all of these problems—regarding development, pollution, oysters, and so on—unfolding well before 2010, it became clear a little more than midway through the first decade of the millennium that the states were going to fall short of most of the targets in the Chesapeake 2000 agreement. But for years, the lack of progress was obscured by overly optimistic reporting by the Chesapeake Bay Program. The Annapolis-based Bay Program is a collaborative, science-based effort of the regional states, the EPA, local universities, and nonprofits. Designed during the Reagan administration, the Bay Program emphasizes states' rights, not top-down federal authority, and in fact has no power to enforce environmental laws or reduce pollution. (This makes the EPA Bay Program different from the EPA itself, whose Region 3 office in Philadelphia has the legal responsibility to enforce the federal Clean Water Act in most of the Chesapeake region.) The most important responsibility of the Bay Program is to provide accurate information about the bay. However, an October 2005 report of the US Government Accountability Office (GAO) concluded that the Bay Program's reports on the bay's health were misleading and "lack credibility" in part because they inappropriately mixed actual water monitoring data with computer modeling of what Bay Program scientists thought theoretically should be happening in the water.[35] "As a result, the public cannot easily determine whether the health of the bay is improving or not," the GAO report stated. "Moreover, the lack of independence in the Bay Program's reporting process has led to negative trends being downplayed and a rosier picture of the bay's health being reported than may have been warranted."

A *Washington Post* investigation reached a similar conclusion in 2008 but also described a financial incentive for the rosy picture of the bay, reporting that "government administrators in charge of an almost $6 billion cleanup of the Chesapeake Bay tried to conceal for years that their effort was failing—even issuing reports overstating their progress—to preserve the flow of federal and state money to the project."[36] Pulitzer Prize–winning reporter David Fahrenthold quoted the Bay Program's director from 1991 to 2001, William Matuszeski, as saying, "To protect appropriations you were getting, you had to show progress. So I think we had to overstate our progress."

In reaction to the GAO report, the EPA Bay Program made several changes, according to Jon Capacasa, the EPA's Region 3 water program director, who oversees the Chesapeake region.[37] "We are certainly doing a better job of labeling information products that go out to the public about whether it's monitored information, actual raw monitoring data, or whether it's simulation models, so that people can understand the difference," Capacasa said. "We have separated our annual reports of progress about the bay to report programmatic activities in one report, and modeled simulation in another report. So we are making a very concerted effort to make it clear to the public which is actual conditions in the bay versus which is predicted or forecast using models." (This is not always true, however, as will be explained later in this chapter.)

During this same period of time, in 2005, the Bay Program changed its definition of low-oxygen "dead zones." "Dead zones" were all over the headlines in the early 2000s, as newspapers reported that nitrogen and phosphorus pollution created algal blooms and low-oxygen (or hypoxic) conditions that smothered 40 percent of the Chesapeake in July 2004.[38]

This number was scandalous because it represented almost half of the bay. The figure *40 percent* was repeated in numerous newspaper articles, books, and even a movie about the bay. Then in 2005, the Bay Program changed its definition of "hypoxic" from less than 5 milligrams per liter of dissolved oxygen to a sliding scale depending on the depth of the water and the type of organism exposed to the water.[39] The EPA Bay Program argued that this change more accurately represented the amount of oxygen that most bottom-dwelling species actually need. The program's lead scientist also suggested that the program did not want industry filing lawsuits against the government for enforcing water quality standards that were unrealistically strict.[40] The Bay Program then redefined "hypoxic" again to mean less than 2 milligrams per liter of dissolved oxygen.[41] The end result of all the redefinition was fewer headlines about giant "dead zones" consuming almost half the bay—not because the problem had disappeared, but because the definition of "dead zones" had changed. Using a more consistent standard of measurement, a 2011 scientific journal article in *Estuaries and Coasts* by researchers at Johns Hopkins and the University of Maryland found "significant increases" in low-oxygen zones (hypoxia) in the bay in early summer over recent decades, perhaps due to climate change, but a "slight decrease" in late-summer hypoxia, caused by decreased nitrogen pollution in the bay.[42]

The goals for the Chesapeake Bay cleanup effort shifted again in 2010 when the EPA imposed a new system of pollution limits. This new system was the Chesapeake Bay "Total Maximum Daily Load" or TMDL. The point of this "historic and comprehensive 'pollution diet'" was that it would impose "rigorous accountability measures" for all of the states in the bay watershed which the EPA could use to penalize states if they failed to meet their targets, according to the TMDL document.[43] The potential penalties include the possible loss of EPA grants for states that don't meet the goals and the federal agency's stepped-up enforcement of existing clean water laws. While the 1987 and 2000 bay agreements had set goals of reducing nitrogen and phosphorus pollution by 40 percent by specific end dates, the 2010 agreement established more moderate targets—a 25 percent cut in nitrogen, a 24 percent reduction in phosphorus, and a 20 percent drop in sediment—with no specified end date by which the numbers must be achieved.[44] Instead, the new agreement uses careful language to avoid any hard deadline for the numeric pollution reductions. The agreement says, "The TMDL is designed to ensure that all pollution control measures needed to fully restore the Bay and its tidal rivers are in place by 2025."[45] Translation: by the year 2025, the bay's water could be just as murky as it is today, but the states and the EPA could declare success within the TMDL framework by claiming that they now have the measures in place (the environmental laws, regulations, and programs) which they believe will eventually, *someday*, restore the health of the estuary. The Bay TMDL boasts of its "rigorous accountability measures," but in this lack of a deadline to achieve numeric pollution targets, the TMDL in some ways demands *less accountability* than either the 1987 or 2000 bay agreements.

Another weakness of the TMDL is that the EPA does not use as its main yardstick for progress toward the 25 percent pollution reduction goal the condition of the bay itself, as reflected by monitoring of the amount of pollution in the water. Instead, the main measurements of success are computer simulations and estimates of how much pollution is *likely* flowing into the bay every year, which are "corrected" and "normalized" for bad weather and include projections of future anticipated reductions.[46] These computer models are driven largely by formulas that attempt to guess how much pollution should, in theory, have been reduced by actions on the ground. For example, if a farmer plants a row of trees along a stream running through his cornfields, the Bay Program model would show in its computer simulations that the stream nearby and bay miles away would instantly get cleaner. However, in reality, it might take decades for the trees to grow to such an extent that they would effectively filter the runoff of fertilizer from the cornfield, and many more years before the stream's health actually improved (if it ever did, at least to the extent predicted by the computer modeling). There is nothing wrong with creating a system to estimate the likely future impact of specific pollution control projects. But such future estimates should not be communicated to the public in a way that implies that they reflect the current condition of the bay, or even the amount of pollution actually entering the bay today, which is what the EPA Bay Program has been doing. This is misleading and has been a cause of overly optimistic assessments of the bay.

An additional level of inaccuracy grows from the fact that the Bay Program model simulation corrects its pollution numbers for the amount of rainfall in a year, dampening out the extremes and instead concentrating on an average or "normal" year. But the reality is that normal is no longer normal. Extreme weather is the new normal. Big storms are becoming increasingly common and more damaging to the bay as climate change continues. To make matters worse, the reservoir behind the Cono-wingo Dam on the Susquehanna River just north of the bay has filled up with sediment. Instead of trapping that pollution from Pennsylvania farm runoff, as it has for decades, the dam—during large rainstorms—now is overwhelmed and allows growing tonnages of sediment laced with phos-phorus to slip into the bay. Climate change could bring even more large storms, releasing even more muck from the Pennsylvania farms.

The result is that there is a large gap between the Bay Program's "nor-malized" computer simulations that the EPA uses to gauge the TMDL's success and the reality in the water. This gap can be seen most vividly by contrasting the Bay Program's computer *modeling* of how much pollution is pouring into the bay with actual water quality *monitoring* of pollutant levels in the water which are analyzed every year by the University of Maryland Center for Environmental Science (UMCES). According to the EPA Bay Program simulations, the amount of phosphorus pollution pour-ing into the estuary declined by an impressive 20 percent between 2009 and 2015, dropping from 19 million pounds in 2009 to 15 million pounds in 2015.[47] This is an upbeat portrayal of reality—not unlike the overly rosy picture for which the GAO criticized the EPA Bay Program in 2005. The University of Maryland scientists use a different approach, examin-ing measurements of the actual amount of phosphorus in the bay itself at 159 sampling stations across the estuary, checked eight times a year from April through October. The UMCES found that the amount of phospho-rus in the bay actually rose between 2009 and 2015, with the bay's health rating for phosphorus falling from a 73 rating out of 100 in 2009 to a 70 in 2015 (with a lower number meaning more phosphorus pollution).[48]

During an interview for my *Environment in Focus* public radio pro-gram, I asked the EPA's Jon Capacasa why his agency didn't use actual water quality monitoring to judge the progress of the Bay TMDL instead of computer simulations, which are less accurate. "We do use monitoring data to gauge the overall direction of things," Capacasa replied.[49] "But the main goal, established by the Bay Program partnership, is one based upon a simulation model to project how the bay will do in 2025 under different conditions."

He explained that the EPA Bay Program corrects for extreme weather events and uses computer simulations with the intention of being more accurate about long-term trends. "We do it so we can compare year to year, with a more apples to apples comparison, rather than comparing a year that has tropical storms happening versus a year that is in drought. Those are really not very convenient for planning purposes," Capacasa said.

I asked him why there are such stark discrepancies between the EPA's computer modeling of phosphorus pollution entering the bay every year and the reality in the water, as demonstrated by the UMCES's data from monitoring. Capacasa admitted that the EPA could have "some missed signals in terms of how we're measuring progress in phosphorus." This inaccuracy, he suggested, could be related to miscalculations of the amount of phosphorus-saturated soils being washed by rain off of farm fields.

"New information like the University of Maryland has shared with us certainly will be considered," Capacasa said.[50] "If our estimates are too rosy, then we certainly want to correct those back to what the data is showing, what the science is showing. We are very open to assessing all of that information."

Nicholas DiPasquale, director of the EPA Chesapeake Bay Program office, said that the Bay Program does not rely solely on computer modeling. He said that the program also uses real water quality monitoring as a way of checking that the bay region states eventually meet water quality standards and restore the estuary to health. Disputing my analogy of referees moving goalposts during a football game, DiPasquale said bluntly, "We don't play games."[51]

Richard Batiuk, associate director for science at the EPA Bay Program, admitted that the goals for the bay cleanup have shifted over the decades. But he described this as a good thing—adaptive management, which keeps changing as better information becomes available.[52] Batiuk also noted that it is the bay region state governments—not just the EPA—that established the bay cleanup goals and then later agreed to adjust the goalposts. "I can't move the goalposts," Batiuk said, emphasizing the role of the states. He added more broadly, "We are always driving toward water quality as our goal."

The good news is that—so far—most of the bay's health indicators have been moving in the right direction since the EPA in 2010 imposed its new system of pollution limits and threats of penalties to states that do not comply. And this improvement has been documented by actual water quality monitoring (as analyzed by the UMCES) and not just estimated by the Bay Program's computer simulations. According to the monitoring, since 2011 the bay has benefited from more underwater grasses, dissolved oxygen, and bottom-dwelling critters like oysters, clams, and worms and has suffered from slightly less nitrogen pollution and algae.[53] As explained in an earlier chapter, however, this positive trend is modest and fragile and could be due in part to recent favorable weather conditions, which could easily reverse themselves. The causes of the improvements likely include upgrades to sewage treatment plants funded by former Maryland governor Robert Ehrlich's 2004 "flush tax" bill and similar investments by Virginia, as well as reductions in air pollution required by a 1990 federal law and federal air regulations imposed by the Obama administration.

To return to the football analogy that opened this chapter, the states and the EPA have moved the goalposts a few times during the course of the Chesapeake Bay cleanup. Now, finally, the ball appears to be moving, too. Whether this slightly improved field position will result in a "win" for the nation's largest estuary depends on how we define victory.

Conclusion
THE FUTURE OF THE BAY

I WAS PADDLING down the Monocacy River in central Maryland on a cool fall afternoon, watching leaves drift down into the rocky shallows, when a bald eagle flew from a branch over the river. At first, I was startled—by its closeness to me as it launched, and by its sudden and powerful wing strokes and massive silvery head. But then I thought to myself, "Oh, it's just a bald eagle. They're everywhere these days." My mental shrug—the "so what" I said to myself—made me reflect. All too often, we take environmental progress for granted. A half century ago, our nation's symbol was nearing extinction. It would be as dead as the dodo today if it were not for the EPA's action to ban the pesticide DDT in 1972, which was strongly opposed by the chemical and agricultural industries. Because of this controversial decision, many fish-eating birds that define the skies of the Chesapeake region came roaring back—including not only eagles but also osprey and great blue heron. These birds, which were almost eradicated, are now so abundant that their piercing cries and rasping croaks are once again a normal part of the background noise around the bay, so common as to be easily ignored.

There are other incredible environmental success stories all around us that we frequently paddle right past in our daily lives. For example, after spotting that eagle, I continued down the Monocacy River, slipped beneath a stone aqueduct for the Chesapeake & Ohio canal, and headed out onto the Potomac. There, a blanket of yellow flowers, atop water stargrass (*Heteranthera dubia*), stretched in a colorful sheet, with fish darting and delving into the clear valleys between mountainous green tufts. Back in the 1960s and 1970s, the Potomac River was almost devoid of life—a reeking toilet that improved dramatically when the federal government forced sewage treatment plants in Washington, DC, and elsewhere to modernize. The river still has a long way to go, of course, as demonstrated by the occasional blobs of blue-green algae and discarded tires in the shallows. But as with the return of the bald eagle, the resurgence of the Potomac River should not be shrugged off. Likewise, we should appreciate the return of great egrets, snowy egrets, Canada geese, and dozens of other migratory bird species that were once blasted from the sky by hunters but came back because of protections in the 1918 Migratory Bird Treaty Act. And we should be thankful for the decline in smog, soot, and mercury air pollution over the past quarter century, driven largely by the federal Clean Air Act of 1970 and its amendments in 1990. All of these victories had one thing in common: they were the result of government action to ban or regulate destructive human behavior. They were not the children of consensus or voluntary partnerships. Ironically, however, in recent years, it is the voluntary, nonregulatory policies favored by industry that have become more fashionable among environmental policy elites than the old-school rules that actually worked.

The salvation of the bald eagle—the icon of America—should be a reminder of how real environmental progress is made. The EPA's decision to ban DDT in the United States was based on incomplete scientific evidence, was attacked by the chemical industry, and is still demonized by some elected officials today (often by the same "conservatives" who take the radical position of denying the science on climate change). Despite the gaps in information and the strident opposition, the blocking of DDT was among the most spectacular environmental victories in history.[1] The lesson is this: we should act decisively and aggressively to protect our environment using the precautionary principle, with the best available evidence, understanding that special interest groups and reflexively anti-government politicians will always try to frustrate action, no matter how necessary it is for the common good.

In response to my argument, some people may counter that DDT is a poor example, because it presented a rare moment of clarity: a single chemical that, by itself, was responsible for the decline of several species of birds. However, I believe that there are numerous other equally clear examples that deserve the same moral conviction and action today, with the understanding that the science on DDT was not actually all that clear back in the 1970s. In twenty-first-century America, cities and towns should not be allowed to intentionally dump raw human feces into waterways that people use for fishing and boating. And yet, Baltimore, Harrisburg, and cities up and down the Susquehanna River and elsewhere continue to deliberately release untreated sewage into Chesapeake Bay tributaries, with only minor repercussions and court consent decrees that allow decades of delay before city officials finally get serious about solving the problem. This is a disturbing failure of law enforcement—as well as an example of our government's chronic neglect of urban infrastructure. To give another example, with the Chesapeake region's landscape being overwhelmed by suburban sprawl, it is nonsensical that thousands of houses remain vacant in Baltimore and other urban areas while fields and forests are being bulldozed for yet more far-flung housing. Our state and federal governments need to institute some commonsense reforms to focus investments (and the approval of building permits) so that our great cities are not abandoned while we blacktop the Chesapeake's countryside. The EPA should not allow the agricultural industry to continue its spraying of pesticides and herbicides, especially glyphosate and neonicotinoids, that have been linked to the mass die-offs of butterflies, bees, and other pollinators. A few percentage points more production from a cornfield is not an acceptable trade-off for the eradication of the monarch butterfly from the American landscape. Coal-fired power plants should not be allowed to taint our fish with mercury and flood our cities with sea-level rise when cleaner sources of electricity are available. Each of these cases, in my mind, is at least as clear as it was with DDT and bald eagles.

There is, of course, an understandable and even desirable conflict between the need to regulate and the resistance of people to regulation. After all, America was founded by folks for whom British tax rules were not their cup of tea. But today's Trump Republicans have thrown themselves overboard in their extreme view that government is the problem, not the solution (to borrow Reagan's formulation). Since the Reagan era, too many policy makers and even environmental organizations have been cowed by Reagan's cowboy rhetoric and have lost the courage to fight for strong government action. The result is that, over the past quarter century, we have launched a flotilla of Chesapeake Bay "restoration" projects that look good at the dock but then sink into the reality of doing nothing to actually reduce pollution. A classic example that many Marylanders see every morning on their drive to work is the "Treasure the Chesapeake" license plate. The millions of dollars raised from these tags are used not to fix leaky sewer systems or to upgrade septic tanks or other projects that improve water quality, but for things—like school field trips—that have more public relations value than pollution control power. As another example, local governments are wasting hundreds of millions of dollars on stream "restoration" projects—bulldozing artificial "S" curves into creeks and building little pools of water with boulders—that have mainly cosmetic value, instead of taking the more difficult step of requiring upstream actions (like mandating that parking lots be built so that they absorb rainwater) that would reduce the runoff before it pollutes the streams.

The argument can even be made that some of these cosmetic and voluntary bay restoration efforts of the past two decades have been counterproductive. Why? Because, politically, they have been used to distract from—and substitute for—the regulations and enforcement that are really needed. Instead of campaigning for strong new laws to fill gaps in the Chesapeake Bay cleanup plan, state officials and even some environmentalists advocate for pollution credit trading as a lower-cost, flexible alternative that offers voluntary participation instead of mandates. Joshua Galperin, director of the Yale Environmental Protection Clinic, describes this trend as "desperate environmentalism."[2] It's an increasingly common but ineffectual response to the rising Republican backlash against the strong environmental laws passed in the 1970s. As Galperin argues,

> Resistance to enforceable environmental protection is rampant. Seeking any conceivable path forward, many young leaders ... turn away from enforceability-based approaches and promote more conservative techniques that they hope will impress and persuade reticent and cynical policymakers and power brokers. If this is environmentalism at all, it is "desperate environmentalism," characterized not by awe, enthusiasm and enjoyment of nature but by appeasement. It relies on utilitarian efficiencies, cost-benefit analyses, private sector indulgences and anthropocentric divvying of natural resources. It champions voluntary commitments, tweaks to corporate supply chains, protection not of the last great places on Earth but of those places that yield profit or services. From market-friendly cap-and-trade to profit-driven corporate social responsibility, desperate environmentalists angle for the least-bad of the worst options rather than the robust and enforceable safeguards that once defined the movement.

Appeasement leads in only one direction: to defeat. As it turns out, the bay's biggest problem is not poultry manure or even human waste: it's hogwash like voluntary partnerships with polluters, laws riddled with loopholes, and pollution trading. We are witnessing an environmental establishment in the Chesapeake region today that—when faced with right-wing obstructionism—all too often surrenders its core values in pursuit of the fundable appearance of collaboration and incremental progress. There are many inspiring exceptions, of course—including heroes like Bonnie Bick, Fred Tutman, Harry Hughes, and John Griffin, who are profiled in this book and whose virtues will live on forever in the Chesapeake.

What do these heroes represent? The idea that we have a fundamental right to clean water and that the Chesapeake Bay is not simply a "resource" for people to use to make money. The bay should be protected because it has inherent value as an ecosystem and as a sacred place of beauty and life. If you believe this, then you also must support the kind of strong and effective environmental laws that have a track record of success. The mandates of the 1972 federal Clean Water Act, as well as taxpayer funding from Maryland's 2004 Bay Restoration Fund Act (the "flush tax" law) and similar investments in Virginia, have modernized many sewage treatment plants in the Chesapeake Bay region and significantly improved water quality in several major rivers, including the Potomac, James, Patuxent, Corsica, and Back Rivers.[3] The federal Clean Air Act amendments of 1990 and other federal and state regulations to reduce air pollution have also worked extraordinarily well to cut nitrogen in streams and rivers that feed the bay. More robust enforcement of environmental laws is still needed, as illustrated by the failures with Baltimore's illegal sewage dumping and Pennsylvania's refusal to require its livestock farms to obtain and follow manure management plans. Moreover, we need more honest assessments of progress in cleaning up the bay, based on real water quality monitoring and not computer simulations and continually shifting goals (see page 187). The toughest issue of all may be climate change. In the end, if we don't find global solutions to this problem, many of our historic Chesapeake communities, like Tangier Island and Hampton Roads, will be under water.

The counterargument to environmental action is always the claim that regulations "kill jobs," but there is no evidence to support this smear.[4] According to an analysis by economist Eban Goodstein and a report from the US Bureau of Labor Statistics, environmental regulations are responsible for less than two-tenths of 1 percent of all layoffs in the United States, with job losses caused much more often by technological advances, globalization, corporate buyouts, and other factors.[5] A good example of crying wolf over environmental rules happened in Maryland in 2006, when Governor Robert Ehrlich warned that an air pollution control bill called the Maryland Healthy Air Act would "dramatically increase the costs of electricity to customers, force at least one power plant to close, and potentially cause rolling blackouts across Maryland."[6] None of these dire warnings proved true. In fact, electricity rates fell (because natural gas prices dropped), and the two largest power companies in the state hired 2,800 construction workers and 92 full-time employees to build and run new air pollution control systems mandated by the law.[7] The past four decades in the United States are full of these types of examples, in which environmental investments require the hiring of people—not the opposite. For example, when the EPA mandates improvements for sewage treatment plants or sewer lines, billions of dollars flow to American contracting and construction businesses. It is true that some coal-fired power plants across the United States are closing. But contrary to what people often read in the media, the closures are not because of environmental regulations but because of technological innovation and competition.[8] The rise of hydraulic fracturing and horizontal drilling has made natural gas so much cheaper than coal that utilities are switching to generating electricity by burning gas. Solar power and wind power are also becoming more efficient and are more effectively competing with coal.

For the Chesapeake Bay cleanup effort, the biggest remaining challenge is runoff pollution from agriculture and suburbia. This is a difficult issue to tackle because the federal Clean Water Act provides much stronger legal authority for controlling discharges from pipes than for controlling the diffuse runoff in rainwater, which is hard even to measure, let alone stop.

However, the good news is that, after decades of no improvement to the bay's overall health, the estuary has shown encouraging signs over the past six years. Although some of the improvement could have been caused by favorable weather conditions, air and water pollution control laws and regulations imposed in the 1990s and 2000s appear to have driven a rise in the overall health of the bay from a 40 rating out of 100 in 2011 to a 54 in 2016, according to the University of Maryland Center for Environmental Science's annual analysis of water quality monitoring.[9] Nitrogen pollution in the bay declined from 2011 to 2016, while algal blooms diminished and levels of dissolved oxygen rose.[10] The estimated oyster population in the bay more than doubled between 2010 and 2014, likely aided by Governor Martin O'Malley's creation of new oyster sanctuaries, before falling off slightly in 2015.[11] Even more impressively, underwater grasses have flourished in recent years, spreading from 79,675 acres in 2010 to 97,433 acres in 2016, the largest amount of aquatic vegetation in the bay since monitoring began more than three decades ago.[12]

To throw some cold water on these positive indicators, however, we need to grapple with the troubling reality that the bay remains significantly murkier today than it was even in the 1980s. Water clarity in the bay (which is often influenced by algal growth and sediment) improved between 2011 and 2016, but it still—with a dismal rating of 24 out of 100 in 2016—remains far lower than it was even in 1986, when it had a 41 rating in the University of Maryland's annual report card.[13]

Overall, the bay appears to have made some progress under the EPA's new system, established in 2010, of pollution limits and potential penalties for Chesapeake states, also known as the bay pollution "diet" or "Total Maximum Daily Load" (TMDL). Some of this improvement was likely caused by a short-term decline in rainfall (which will not likely continue) and by regulations and laws passed years or decades earlier that required time to produce results. But it would be irrational to dismiss as a coincidence the timing of the bay's resurgence and the leadership role that the EPA assumed in 2010 by imposing the new federal pollution limits. Clearly, we are seeing some meaningful, although modest, new success. However, this forward momentum is threatened by President Trump's draconian cuts to the EPA and the Republican Party's fanatical, fact-blind anti-regulatory agenda. Instead of cuts and deregulation, what the bay desperately needs is stronger rules and escalated enforcement.

Eventually, the Trumpian wave will pass like a fever through the body politic and subside. Assuming that America survives and sanity returns, progress on the bay may resume. Below are 10 steps that—although admittedly very ambitious—would, under the right political conditions, propel the bay restoration forward under full sail. Although the Chesapeake cleanup is often described as a financial burden on taxpayers, the following ideas would not require a penny from taxpayers because they would compel polluters to pay, or shift money between existing government programs.[14]

1. To reduce the biggest single source of pollution in the bay, the US Department of Agriculture should require any farmers who receive federal taxpayer subsidies—as a condition for pocketing those entitlements—to follow best management practices to prevent the runoff of fertilizer and sediment into nearby streams.[15] The mandatory actions for farmers should include fencing their cattle out of waterways and maintaining wide buffer strips of grasses or trees as filters along streams, where the spreading of fertilizer should be prohibited.

2. After a warning, Pennsylvania should impose fines on all farmers who fail to comply with the state's Clean Streams Law. The law has required livestock operators to obtain and follow manure management plans for more than three decades. Disturbingly, however, a large portion of the livestock farms in the bay watershed of Pennsylvania still do not have or follow these plans, meaning that they dump or spread large amounts of manure that runs off into adjacent waterways, creating algal blooms and "dead zones" in the Chesapeake.[16] This is an insult to all who love the bay or the rule of law, and it must stop.

3. The bay region states should not approve permits for any more factory-sized poultry houses or livestock operations until the big meat companies start paying for more responsible disposal systems for the vast amount of excess manure produced by their contract farmers. Companies like Perdue and Tyson (not taxpayers) should pick up the tab for burying the excess waste in lined landfills, or shipping the manure to other parts of the country that have low levels of phosphorus in their soil.

4. Maryland and Virginia should ban all harvesting of wild oysters in the Chesapeake Bay until populations of the shellfish—now at perhaps 1 percent of historic levels—can recover to at least 25 percent of their original numbers. Without decisive action like this, the "great shellfish bay" will soon have no shellfish.

5. The bay region states should prohibit the sale and use of lawn fertilizer. While some fertilizer is necessary for crops, turfgrass lawns are purely cosmetic, and most of them do not need these chemicals, which are like rocket fuel for algal blooms in the bay.

6. To curb suburban sprawl, states and counties should not approve any more building permits for new subdivisions on farmland or in the countryside until the abandoned homes in nearby cities and towns are occupied. Developers who request permits to build in fields and forests should be directed to first recycle the tens of thousands of empty houses in historic communities like Baltimore, Crisfield, Cambridge, and Richmond.

7. Any developers who want to build new parking lots in suburban or rural areas should be required to use permeable pavement or gravel that absorbs rainwater and reduces runoff pollution into streams and the bay. This additional cost and hassle will discourage sprawl and encourage investment in urban areas.

8. All new homes built in the Chesapeake Bay watershed that are not connected to state-of-the-art sewage treatment plants should be required to install septic systems (underground waste tanks) with the best available technology for filtering out nitrogen pollution. Homes on septic tanks leak up to 10 times more nitrogen pollution into the bay than homes connected to modern sewage plants.[17]

9. To reduce the cans, bottles, and bags that trash even the bay's most remote shorelines and islands, the bay region states should impose 10 cent deposits on the purchase of these drink containers, as well as 5 cent fees on plastic bags, with the money refunded when people return them to stores or recycling centers. Fast-food containers made of foam and other materials that do not biodegrade should be banned.

10. The federal government should slash US military spending by one-third, leaving forces that would remain, by far, the most powerful in the world. The $200 billion saved each year should be invested in America instead of being used to defend the interests of wealthy foreign countries such as Saudi Arabia and South Korea. Much of the excess funds should be directed to rebuild infrastructure in the United States, especially sewer and stormwater pollution control systems, and to upgrade sewage treatment plants. This reinvestment in cleaning up America's waterways would create millions of private-sector jobs in local engineering and construction firms that could not be outsourced to China, India, or Mexico.

These ideas (many of which have been rattling around for years) would not only help restore the Chesapeake Bay but also boost the economies of Maryland, Virginia, Pennsylvania, and the other regional states. So why are *none* of these steps—not one!—included in the states' federally approved bay cleanup plans?

Well, in a few cases, as with my recommendations 1 and 10 above, federal action would be required, which would be politically impossible—even laughable—with Congress and the White House controlled as they are today by anti-regulatory radicals determined to spend ever more money on the Pentagon. All federal action today is in the direction of slashing environmental regulations and boosting the budget for the military, not the opposite. But in other cases, the obstacle is state politics. The Chesapeake region states have not been at all equal in their efforts to clean up the nation's largest estuary. In particular, even in the face of strong agricultural and development lobbies, Maryland has moved forward with some highly controversial—and praiseworthy—steps in recent years to improve protections for blue crabs and oysters, upgrade sewage treatment plants, and reduce runoff pollution. Virginia also deserves applause for banning the destructive practice of dredging for hibernating female crabs in the winter and for investing almost a billion dollars to upgrade its sewage plants. By far the biggest problem is Pennsylvania, both because it is the state that contributes the most pollution to the bay and because it has done so little.

So why hasn't the EPA lowered the boom on Pennsylvania for dumping on its downstream neighbors? The federal agency has taken a few modest actions, but, in general, its response has been slow and weak, even under Democrats. This may be in part because the federal Clean Water Act grants the EPA only limited power over agriculture, the main source of pollution from Pennsylvania. And the Republican-led Congress has been far more eager to eviscerate the EPA's funding and power than to expand its authority over farms. Nicholas DiPasquale, director of the EPA Chesapeake Bay Program, described the agency's political position this way (using slightly saltier language): "The Farm Bureau has been bashing the [stuffing] out of us for our so-called 'over-reach.'"[18] In July 2016, the House, led by staunch Farm Bureau ally Robert Goodlatte, a Republican representative from Virginia, voted in favor of an appropriations bill amendment that would strip the EPA of its ability to impose penalties on states that fail to meet their pollution limits under the bay pollution "diet" or TMDL.[19] Within the Chesapeake Bay watershed, a majority of US representatives (by a 20–18 margin) voted in favor of Goodlatte's proposed curbs on the EPA's power, with all 20 of the favorable votes cast by Republicans.[20] This political movement could potentially derail the whole Chesapeake Bay cleanup project. With the election of President Trump—who pledged during his campaign to "get rid of [the EPA] in almost every form"—this threat to the bay became more real.[21] Dark clouds formed over the bay when Trump proposed to eliminate all funding for the EPA Chesapeake Bay Program and selected as his EPA administrator Scott Pruitt, the Oklahoma attorney general. Pruitt, in 2010, filed a legal action in support of the American Farm Bureau's lawsuit to kill the EPA's Chesapeake cleanup plan (the Bay TMDL). Even though this lawsuit failed, Pruitt could achieve the same result by simply doing nothing as EPA administrator. His murder weapon for the Chesapeake could be silence—in other words, the EPA's passive refraining from penalizing states (especially Pennsylvania) that miss their pollution reduction targets. Through EPA inaction, the whole Bay TMDL system would become ineffective. However, the recent anti-regulatory political winds are only part of the problem. For years before Trump's election, the EPA's existing authority allowed it to force Pennsylvania to upgrade its sewage treatment plants to match the levels of Maryland and Virginia. But the federal agency—even under pro-environment, pro-regulatory President Barack Obama—chose not to take even this reasonable step because it would be politically unpopular in Pennsylvania.

Letting Pennsylvania off the hook like this has been unfair to taxpayers in Maryland and Virginia. But the truth is that government agencies like the EPA are indirect reflections of the will of the voters. There may be a perception among government officials that while voters give lip service to the idea of a clean bay, their support is as shallow as the Chesapeake itself. Even in a bay-friendly state like Maryland, the fierce backlash of voters against stormwater pollution control fees in 2012—the "rain tax" that imposed costs of a few dollars per month on households—was a sobering example of how little most voters are willing to sacrifice.

What is going on here? Why are we willing to give so little for a waterway that is the geographical and cultural centerpiece of our state? I have often thought that what we need is a new definition of patriotism, a vision of the "American Dream" that is divorced from its present selfishness: my house, my money, and my rights. Since Reagan's labeling of government as "the problem, not the solution," we have seen a cultural shift away from the idea of spending on any kinds of programs that serve the majority of people instead of the narrow interests of the wealthy and powerful. The result has been a snowballing neglect of public spaces, public infrastructure, and the public good in favor of private property and private profit. We have witnessed a growing hostility toward serving anything beyond business. And in reaction to this, we have seen among liberals the growth of desperate environmentalism, which dismisses the idea of strong environmental laws as impossible.

Let's look deeper beneath the surface. A major problem in America today is that people increasingly walk around inside their own personal "fact" bubbles. These hermetically sealed "realities" are growing ever more separate from real reality because of our increasingly fragmented media world and the ability of people to self-select what truths they are exposed to on cable television or the Internet. Everyone believes that he or she is armed with the truth and is acting virtuously. Developers believe they are creating jobs; farmers believe they are fighting to hold on to their family heritage; watermen believe they need to pay their bills so they don't lose their boats and homes, and regulations get in the way of that. But few people can see outside of their own heads or beyond their own self-interest. As Tangier Island waterman Ooker Eskridge said of "saving the bay" (on page 95), "Most watermen feel like I do: To save the resource, if they eliminate the watermen, it's not worth it. Watermen don't give a rip about the bay if they are going to get us off of the bay."

Watermen don't trust scientists. Farmers don't trust the government. Elected officials don't trust that the public really cares enough to pay attention and vote for their interests. And voters don't trust their elected representatives. The Chesapeake Bay cleanup is, at bottom, a social problem—really, an almost religious crisis—and in the end what we need is more trust and more faith. What we are up against is not just water pollution, but a radical anti-government ideology that worships the personal acquisition of money and power over all else and is fundamentally selfish and destructive not only to nature but to humanity itself. This is an issue far more grave than simply the Chesapeake Bay. We need to save ourselves. In the face of a fierce and cynical wave breaking over us, we need to rebuild a basic confidence in the ability of people to work together through democratic government to make the world a better place. Yes, financial profit is important, as is free enterprise and individual virtue. But we also need all people, as well as all businesses, to act together against their own self-interests for the benefit of our shared world—our waters, our lands, and our health. The problem is that nobody wants to sacrifice for the good of others. The only way to cross that river is by loving your neighbor, and that requires a cleanup of the soul.

Notes

Introduction: The Waves Threatening Chesapeake Country

1. Many of the quotes, facts, and descriptions in the opening scene of this introduction are reprinted with permission from a *Baltimore Sun* article I published on the Blackwater National Wildlife Refuge. Tom Pelton, "Blackwater Preserve's Natural Balancing Act," *Baltimore Sun*, October 2, 2005, http://articles.baltimoresun.com/2005-10-02/news/0510020136_1_blackwater-national-wildlife-wildlife-refuge-wetlands.

2. Ellen K. Silbergeld, *Chickenizing Farms and Food: How Industrial Meat Production Endangers Workers, Animals, and Consumers* (Baltimore: Johns Hopkins University Press, 2016).

3. Chesapeake Bay Program, "Facts and Figures," www.chesapeakebay.net/discover/bay101/facts.

4. Chesapeake Bay Foundation, "Polluted Runoff: How Investing in Runoff Pollution Control Systems Improves the Chesapeake Bay Region's Ecology, Economy, and Health," January 2014, www.cbf.org/document-library/cbf-reports/2014-Polluted-Runoff-Report-compressed20b2.pdf. For farm runoff numbers from the Chesapeake Bay Program, see www.chesapeakebay.net/issues/issue/agriculture#inline.

5. Many of the historical facts in this chapter are from John R. Wennersten, *The Chesapeake: An Environmental Biography* (Baltimore: Maryland Historical Society Press, 2001).

6. University of Maryland Center for Environmental Science (UMCES), Eco-Health Report Card for the Chesapeake Bay, 2016, http://ecoreportcard.org/report-cards/chesapeake-bay/health/. Note: UMCES continually updates and corrects the numbers on the report cards. The numbers in this book are from a July 2, 2017, viewing of the UMCES website and data as of that date. See also US Geological Survey and partners, "New Insights: Science-Based Evidence of Water Quality Improvements, Challenges, and Opportunities in the Chesapeake," March 2014, www.chesapeakebay.net/channel_files/21409/new_insights_report.pdf.

7. Virginia Institute of Marine Science of the College of William & Mary, "SAV in Chesapeake Bay and Coastal Bays," http://web.vims.edu/bio/sav/index.html.

8. UMCES Eco-Health Report Card for the Chesapeake Bay, 2016, http://eco reportcard.org/report-cards/chesapeake-bay/health/.

THE WATERS
Susquehanna River: A Winter's Journey to the Father of the Bay

1. Many of the descriptions and quotes in this chapter are from the *Environment in Focus* radio program "Mysterious Disease in Bass Leaves Fishermen Cold," February 29, 2012, http://programs.wypr.org/podcast/2-29-12-mysterious-disease -bass-leaves-fishermen-cold. Other details and quotes are from a telephone interview with Juan Veruete, April 8, 2015.

2. Chesapeake Bay Program, "Facts and Figures," www.chesapeakebay.net/dis cover/bay101/facts.

3. Darryl Fears, "Will Dredging Alleviate the Conowingo Dam Sediment Issue?," *Washington Post*, April 7, 2015.

4. American Rivers, "America's Most Endangered Rivers of 2005," https:// s3.amazonaws.com/american-rivers-website/wp-content/uploads/2016/02/2422 0916/2005-mer-report.pdf.

American Rivers, "America's Most Endangered Rivers of 2016," http://b.3cdn .net/amrivers/63c3def8e0d30964cc_4btm67ybb.pdf.

5. Susan Q. Stranahan, *Susquehanna: River of Dreams* (Baltimore: Johns Hopkins University Press, 1993).

6. Chesapeake Bay Commission, "Healthy Livestock, Healthy Streams: Policy Actions to Promote Livestock Stream Exclusion," May 2015, www.chesbay.us/Pub lications/Healthy%20Livestock,%20Healthy%20Streams.pdf.

7. Rona Kobell, "Susquehanna Ailing Not Impaired, Pennsylvania Says," *Bay Journal*, July 29, 2016, www.bayjournal.com/article/susquehanna_ailing_but_not _impaired_pennsylvania_says.

Gunpowder River: Development Patterns, as Reflected in the Water

1. Many of the quotes and descriptions in this chapter are from the *Environment in Focus* radio program "Fish Reproduction Failure Linked to Suburban Development," September 4, 2013, http://programs.wypr.org/podcast/9-4-13-fish-repro duction-failure-linked-suburban-development.

2. EPA Chesapeake Bay Program, "Chesapeake Bay Watershed Population," www .chesapeakebay.net/indicators/indicator/chesapeake_bay_watershed_population.

3. Maryland Department of Planning, Population Projections (based on US Census data), www.mdp.state.md.us/msdc/s3_projection.shtml.

4. Chesapeake Bay Foundation, "Polluted Runoff: How Investing in Runoff Pollution Control Systems Improves the Chesapeake Bay Region's Ecology, Economy, and Health," January 2014, www.cbf.org/document-library/cbf-reports/2014-Pol luted-Runoff-Report-compressed20b2.pdf.

5. Ibid.

6. The Interagency Mattawoman Ecosystem Management Task Force, "The Case for Protection of the Watershed Resources of Mattawoman Creek," March 15, 2012, http://dnr2.maryland.gov/fisheries/Documents/Mattawoman_Ecosystem _Final_Report_March_2012.pdf.

Corsica River: A Murky River Experiment's Clear Lessons

1. Chris Guy, "State to Clean Up Corsica River: $20 Million and Research of 30 Agencies Will Aid Shore Tributary," *Baltimore Sun*, September 28, 2005, http:// articles.baltimoresun.com/2005-09-28/news/0509280336_1_corsica-river -tributary-centreville.

2. Many of the quotes and descriptions in this chapter are from the *Environment in Focus* radio program "Murky River Experiment Produces Clear Lessons for the Chesapeake Bay," June 3, 2015, http://wypr.org/post/murky-river-experiment -produces-clear-lessons-chesapeake-bay#stream/0.

3. Numbers are from an interview with Frank DiGialleonardo, founding president of the Corsica River Conservancy, on May 29, 2015. He provided a printout with many of the figures cited in this chapter. Also helpful was a paper by Walter R. Boynton, Jeremy M. Testa, and W. Michael Kemp, "An Ecological Assessment of the Corsica River Estuary and Watershed: Scientific Advice for Future Water Quality Management," University of Maryland Center for Environmental Science, October 30, 2009.

4. Chris Guy, "Md. Probes Allegations at Treatment Plant: Company Running Facility in Centreville Accused of Falsifying Records, Reports," *Baltimore Sun*, October 5, 2004, http://articles.baltimoresun.com/2004-10-05/news/0410050324_1 _centreville-treatment-plant-corsica-river.

5. Water quality data on the Corsica River were provided by the Maryland Department of the Environment via e-mail on April 30, 2015.

6. Ibid.

7. University of Maryland Center for Environmental Science, EcoHealth Report Card on the Chesapeake Bay, 2015, http://ecoreportcard.org/report-cards/ chesapeake-bay/health/.

8. Telephone interview with Kevin Smith, assistant director of the Chesapeake and Coastal Service at the Maryland Department of Natural Resources, May 8, 2015.

Patuxent River: The Patuxent River Will Break Your Heart

1. Many of the quotes and observations in this chapter are from the *Environment in Focus* radio program "Activist Throws His Heart, Lawsuits—and Nearly His Life—into the Patuxent River," July 1, 2015, http://wypr.org/post/activist -throws-his-heart-lawsuits-and-nearly-his-life-patuxent-river-0#stream/0. Fred Tutman serves on the board of the Environmental Integrity Project.

2. Tutman's statements about Chalk Point were made on May 15, 2015. After that, a consent decree settled the case and required the company to reduce its pollution.

3. John R. Wennersten, *The Chesapeake: An Environmental Biography* (Baltimore: Maryland Historical Society, 2001).

4. Robert de Gast, *Five Fair Rivers: Sailing the James, York, Rappahannock, Potomac, and Patuxent* (Baltimore: Johns Hopkins University Press, 1995).

5. Tom Horton, *Bay Country* (Baltimore: Johns Hopkins University Press, 1987).

6. Patuxent Riverkeeper report, "Patuxent River 20/20: The Need for Effective Action and Effective Solutions," December 2007, www.paxriverkeeper.org/prk/file/Pax2020_final-reduced-more.pdf.

7. Richard Hall, Maryland's Secretary of Planning, "The Patuxent River's Unsatisfied Man," *Baltimore Sun*, June 10, 2012.

8. University of Maryland Center for Environmental Science EcoCheck Report Card on the Health of the Chesapeake Bay, http://ecoreportcard.org/report-cards/chesapeake-bay/health/.

9. Ibid.

10. Frederick Tutman, "You Pollute, and Then You Sue?" *Waterkeeper* 13, no. 1 (2016), https://waterkeeper.org/magazine/volume-13-issue-1/you-pollute-and-then-you-sue/.

11. In an e-mail on February 28, 2017, Frederick Tutman said that the attack happened in 2014 in Ocean City at the Annual Maryland Association of Counties meeting and that no charges were filed. The development that Tutman had challenged was the Woodmore Town Center in Largo, Maryland.

Potomac River: Black Roses in the Nation's River

1. Some of the descriptions and observations in this chapter are from the *Environment in Focus* radio program "A Garden of Black Roses Grows in a River," August 28, 2013, http://programs.wypr.org/podcast/8-28-13-garden-black-roses-grows-river.

2. Ben Guarino, "'We've Primed the System': Why Disgusting Toxic Blue-Green Algae Blooms Seem Increasingly Common," *Washington Post*, July 25, 2016, www.washingtonpost.com/news/morning-mix/wp/2016/07/25/weve-primed-the-system-why-toxic-blue-green-algae-blooms-seem-increasingly-common/?wpisrc=nl_mix&wpmm=1.

3. EPA Chesapeake Bay Program, "River Flow into Chesapeake Bay," www.chesapeakebay.net/indicators/indicator/river_flow_into_chesapeake_bay. Note: this length of the Potomac River includes the North Branch.

4. Many of the historical facts in this chapter are from a pair of books: Garrett Peck, *The Potomac River: A History and Guide* (Charleston, SC: History Press, 2012); Robert de Gast, *Five Fair Rivers: Sailing the James, York, Rappahannock, Potomac, and Patuxent* (Baltimore: Johns Hopkins University Press, 1995). More recent facts about the Blue Plains Wastewater Treatment Plant can be found on the website of the District of Columbia Water and Sewer Authority, www.dcwater.com/wastewater/blueplains.cfm. Another good source of information about the Potomac River and Blue Plains are the archives of the *Bay Journal*; see www.bayjournal.com.

James River: The Spirits of the Swamp

1. David A. Price, *Love and Hate in Jamestown: John Smith, Pocahontas, and the Start of a New Nation* (New York: Vintage Books, 2003).

2. Bob Deans, *The River Where America Began: A Journey along the James* (Lanham, MD: Rowman & Littlefield, 2007).

3. Ibid.

4. Interview with Jamie Brunkow, Lower James Riverkeeper, March 27, 2015.

5. US Geological Survey, "Summary of Trends Measured at the Chesapeake Bay Tributary Sites: Water Year 2013 Update," December 5, 2014.

6. James River Association, "State of the James, 2015," https://jrava.org/wp -content/uploads/2016/04/state-of-the-james.jpg.

7. University of Maryland Center for Environmental Science EcoCheck Report Card on the Health of the Chesapeake Bay, http://ecoreportcard.org/report -cards/chesapeake-bay/health/.

8. Chesapeake Bay Foundation, "Polluted Runoff: How Investing in Runoff Pollution Control Systems Improves the Chesapeake Bay Region's Ecology, Economy, and Health," January 2014, www.cbf.org/document-library/cbf-reports/2014-Pol luted-Runoff-Report-compressed20b2.pdf.

Southern Bay: A Kayak Expedition to the Mouth of the Chesapeake

1. Many of the quotes and descriptions in this chapter are from the *Environment in Focus* radio program "A Kayak Expedition to the Mouth of the Chesapeake Bay," June 10, 2015, http://wypr.org/post/kayak-expedition-mouth-chesapeake-bay #stream/0.

2. Much of the history in this chapter is drawn from David A. Price, *Love and Hate in Jamestown: John Smith, Pocahontas, and the Start of a New Nation* (New York: Vintage Books, 2003). Also important was Bob Deans, *The River Where America Began: A Journey along the James* (New York: Rowman & Littlefield, 2007); and Eric Mills, *Chesapeake Bay in the Civil War* (Centreville, MD: Tidewater, 1996). In addition, details came from the Virginia Foundation for the Humanities' guide, *Encyclopedia Virginia*; see www.encyclopediavirginia.org/.

3. University of Maryland Center for Environmental Science EcoCheck Report Card on the Health of the Chesapeake Bay, http://ecoreportcard.org/report -cards/chesapeake-bay/health/.

4. Ibid.

5. Ibid.

6. Ibid.

7. Maryland Department of Natural Resources press release, "Maryland's Oyster Population Continues to Improve, Highest since 1985," May 7, 2014, http:// news.maryland.gov/dnr/2014/05/07/marylands-oyster-population-continues -to-improve-highest-since-1985/.

8. Virginia Institute of Marine Science of the College of William & Mary, "SAV in Chesapeake Bay and Coastal Bays," http://web.vims.edu/bio/sav/index.html.

9. Keith N. Eshleman and Robert D. Sabo, "Declining Nitrate-N Yields in the Upper Potomac River Basin: What Is Really Driving Progress under the Chesapeake Bay Restoration?," *Atmospheric Environment* 146 (2016): 280–89.

THE PEOPLE

Harry Hughes: The Unexpected Captain of the Bay Cleanup

1. Interviews with former Maryland governor Harry Hughes, March 6 and 7, 2015.

2. Many of the historical facts in this chapter are from Governor Harry Hughes's autobiography, Harry Hughes with John W. Frece, *My Unexpected Journey: The Autobiography of Governor Harry Roe Hughes* (Charleston, SC: History Press, 2006).

3. Karl Blankenship, "Charles 'Mac' Mathias, Founder of Bay Cleanup Effort, Dies," *Bay Journal*, March 1, 2010, www.bayjournal.com/article/charles_mac_mathias _founder_of_bay_cleanup_effort_dies.

4. Hughes with Frece, *My Unexpected Journey*.

5. Atlantic States Marine Fisheries Commission, Atlantic Striped Bass, 2013 Stock Status, www.asmfc.org/species/atlantic-striped-bass.

6. Howard R. Ernst, *Fight for the Bay: Why a Dark Green Environmental Awakening Is Needed to Save the Chesapeake Bay* (New York: Rowman & Littlefield, 2010).

Parris Glendening: The Green Governor and the Cell from Hell

1. Interview with former governor Parris Glendening, April 9, 2015.

2. Rebecca Lewis, Gerrit-Jan Knaap, and Jungyul Sohn, "Managing Growth with Priority Funding Areas: A Good Idea Whose Time Has Yet to Come," *Journal of the American Planning Association* 75, no. 4 (October 1, 2009): 457–78, http://dx .doi.org/10.1080/01944360903192560.

3. Interview with former Maryland state senator Gerald Winegrad, April 1, 2015.

4. Report of the Governor's Blue Ribbon Citizens Pfiesteria Piscicida Action Commission, November 3, 1997, http://msa.maryland.gov/megafile/msa/speccol /sc5300/sc5339/000113/000000/000152/unrestricted/20040010e.html.

5. Ted Shelsby, "Grain Farmers Honor Protector Guns, Cecil Delegate, Helped Soften Impact of Pollution Control," *Baltimore Sun*, July 14, 1998, http://articles. baltimoresun.com/1998-07-24/business/1998205140_1_maryland-farmers -grain-producers-eastern-shore.

6. Daniel LeDuc and Peter S. Goodman, "Governor Seeks Controls to Protect Bay," *Washington Post*, January 22, 1998, www.washingtonpost.com/wp-srv/local /longterm/library/pfiesteria/controls0122.htm.

7. Environmental Integrity Project, "Poultry's Phosphorus Problem: Phospho-

rus and Algae in Eastern Shore Waterways: High Concentrations, No Improvement in Past Decade," July 2014, www.environmentalintegrity.org/wp-content/uploads /2016/11/2014-07_Poultrys_Phosphorus_Problem.pdf.

8. Michelle Perez, "Regulating Farmer Nutrient Management: A Three-State Case Study on the Delmarva Peninsula," *Journal of Environmental Quality* 44, no. 2 (March 2015): 402–14, https://dl.sciencesocieties.org/publications/jeq/pdfs/44 /2/402.

9. Environmental Integrity Project, "Manure Overload on Maryland's Eastern Shore: Phosphorus Regulations Are Needed to Reduce Poultry Pollution into the Chesapeake Bay," December 2014, www.environmentalintegrity.org/wp-content /uploads/2014/12/POULTRY-REPORT.pdf.

10. Salisbury University Business Economic and Community Outreach Network (BEACON) study for the Maryland Department of Agriculture, "A Scenario Analysis of the Potential Costs of Implementing the Phosphorus Management Tool on the Eastern Shore of Maryland," November 2014, http://mda.maryland .gov/Documents/pmt-analysis.pdf.

11. Maryland Department of Agriculture Nutrient Management Program, *Fiscal Year 2015 Annual Report*, http://mda.maryland.gov/resource_conservation /counties/NMPAnnualReport%202015FINAL_web.pdf.

12. Perez, "Regulating Farmer Nutrient Management."

13. Ibid.

14. Maryland Farm Bureau press release, "Good News on Bay Clean Up Effort, but Out-of State Environmental Activists Can't Stop Attacking," July 15, 2014, http://mdfarmbureau.com/wp-content/uploads/2014/07/GoodBayNews.pdf.

15. University of Maryland Center for Environmental Sciences EcoCheck Health Report Card on the Chesapeake Bay, http://ecoreportcard.org/report -cards/chesapeake-bay/health/. See also US Geological Survey, "Understanding Nutrients in the Chesapeake Bay Watershed and Implications for Management and Restoration—the Eastern Shore," Circular 1406, February 2015, https://pubs.er .usgs.gov/publication/cir1406.

16. Scott Phillips, US Geological Survey Chesapeake Bay Coordinator, "Testimony for Maryland General Assembly Hearings on the Phosphorus Management Tool," presented to the Maryland Senate Environment, Health and Education Committee and House Environment and Transportation Committee on February 23, 2015.

17. A. R. Place, K. Saito, J. R. Deeds, J. A. F. Robledo, and G. R. Vasta, "A Decade of Research on Pfiesteria Spp. and Their Toxins: Unresolved Questions and an Alternative Hypothesis," in *Seafood and Freshwater Toxins: Pharmacology, Physiology, and Detection, Seafood and Freshwater Toxins, Pharmacology, Physiology, and Detection*, 2nd ed., ed. Luis M . Botana (Boca Raton, FL: CRC, 2008). See also D. E. Terlizzi, "*Pfiesteria* Hysteria, Agriculture, and Water Quality in the Chesapeake Bay: The Extension Bridge over Troubled Waters," *Journal of Extension* 44, no. 5 (2006),

www.joe.org/joe/2006october/a3.php; interview with Wolfgang Vogelbein, biologist with the Virginia Institute of Marine Science, April 2012; interview with Dr. Allen Place at the University of Maryland Center for Environmental Science, May 8, 2015.

John Griffin: Watching Over the Wild

1. Many of the quotes and observations in this chapter also appear in the *Environment in Focus* radio program "Honoring a Champion of the Chesapeake," March 17, 2015, http://wypr.org/post/honoring-champion-chesapeake-bay.

2. Chesapeake Bay Foundation, "Bad Water and the Decline of Blue Crabs," December 2008, www.conservationgateway.org/Documents/CBF-BadWatersReport.pdf.

3. Karl Blankenship, "Blue Crab Fishery in Bay to Get Federal Economic Aid," *Bay Journal*, October 1, 2008, www.bayjournal.com/article/blue_crab_fishery_in_bay_to_get_federal_economic_aid.

4. Chesapeake Bay Program, "Chesapeake Blue Crab Population Reaches Highest Level in Nearly 20 Years," April 19, 2012, www.chesapeakebay.net/blog/post/chesapeake_bay_blue_crab_population_reaches_highest_level_in_nearly_20_year.

5. Chesapeake Bay Program, "Chesapeake Blue Crab Abundance Remains Low," May 7, 2014, www.chesapeakebay.net/blog/post/chesapeake_bays_blue_crab_abundance_remains_low; Maryland Department of Natural Resources, "Chesapeake Bay Blue Crab Population Shows Modest Improvement," April 27, 2015, http://news.maryland.gov/dnr/2015/04/27/chesapeake-bay-blue-crab-population-shows-modest-improvement/.

6. Candus Thomson, "Total of Ten Tons of Illegally Caught Striped Bass Found over Three Days," *Baltimore Sun*, February 2, 2011, http://articles.baltimoresun.com/2011-02-02/sports/bs-sp-rockfish-poaching-0203-20110202_1_illegal-nets-commercial-gill-net-season-larry-simns.

7. Maryland Department of Natural Resources, "Maryland's Oyster Population Continues to Improve, Highest since 1985," May 7, 2014, http://news.maryland.gov/dnr/2014/05/07/marylands-oyster-population-continues-to-improve-highest-since-1985/.

Bonnie Bick: A Soft-Spoken Warrior for the Chesapeake's Forests

1. Many of the quotes, descriptions, and facts in this chapter are from two *Environment in Focus* radio programs, (1) "A Soft Spoken but Fierce Hero of Maryland's Forests and Streams," March 10, 2016, http://wypr.org/post/soft-spoken-fierce-hero-maryland-s-forests-and-streams; and (2) "A Victory by Conservationists—and Its Aftermath," May 2, 2012, http://programs.wypr.org/podcast/5-2-12-victory-conservationists-and-its-aftermath; and Tom Pelton, "Highway Threatens Creek Filled with Life; Connector Could Foul Mattawoman in Southern Maryland," *Baltimore Sun*, April 7, 2008, http://articles.baltimoresun.com/2008-04-07/news/0804070170_1_mattawoman-creek-charles-county-southern-maryland.

2. Paul Schwartzman, "Maryland Activist Won't Give Up Fight—or Home," *Washington Post*, May 3, 2004, www.washingtonpost.com/archive/local/2004/05 /03/md-activist-wont-give-up-fight-or-home/323ff75f-3e2e-40de-8410-304517d 91473/.

3. Roger K. Lewis, "Big Ideas and the Tale of Chapman's Landing in Charles County, Md.," *Washington Post*, October 8, 2015, www.washingtonpost.com/real estate/big-ideas-and-the-tale-of-chapmans-landing-in-charles-county-md/2015 /10/07/56b10138-6c52-11e5-b31c-d80d62b53e28_story.html. Note: $3.2 million for the land preservation came from the Richard King Mellon Foundation thanks to the efforts of the Conservation Fund.

4. Pelton, "Highway Threatens Creek."

5. The Maryland Department of the Environment approved 99.8% (5,873 out of 5,883) of the applications for wetlands destruction permits which builders submitted to the agency in 2009, 2010, and 2011, according to state records. Numbers from state data were obtained through a Maryland Public Information Act request to the Maryland Department of the Environment. Figures were reported in the *Environment in Focus* radio program "The Illusion of Wetlands Restoration," February 8, 2012, http://programs.wypr.org/stationprogram/environment-focus-tom -pelton-2?page=10.

Michael Beer: The Lorax of Baltimore

1. The quotes and observations in this chapter are from the *Environment in Focus* radio program "The Lorax of Baltimore Plants His Last Tree," September 2, 2014, http://wypr.org/post/lorax-baltimore-plants-his-last-tree#stream/0.

Carole Morison: Free as a Bird Now

1. Many of the quotes and facts in this chapter are from the *Environment in Focus* radio program "Farmer Profits by Flying from Big Poultry Company," June 17, 2014, http://wypr.org/post/farmer-profits-flying-big-poultry-company, as well as e-mails from Carole Morison on May 15 and 20, 2016.

2. Letter from Carole Morison to US Attorney General Eric Holder and Department of Agriculture Secretary Tom Vilsack, December 30, 2009, www.justice .gov/sites/default/files/atr/legacy/2010/02/24/255197.pdf.

3. Perdue stopped adding a form of arsenic (Roxarsone) to poultry feed in April, 2007, according to a company press release, "Statement on Roxarsone," www.per duefarms.com/News_Room/Statements_and_Comments/details.asp?id=285&title =Statement%20on%20Roxarsone%20. The company's decision came just weeks after a front-page *Baltimore Sun* article detailed Carole Morison's feeding of her chickens arsenic in Perdue poultry feed and her concerns about its potential health impact. Tom Pelton, "Arsenic's Use in Chicken Feed Troubles Health Advocates," *Baltimore Sun*, March 10, 2007, http://articles.baltimoresun.com/2007-03-10/news /0703100106_1_organic-arsenic-chicken-feed-feed-additive.

4. Ellen K. Silbergeld, *Chickenizing Farms and Food: How Industrial Meat Production Endangers Workers, Animals, and Consumers* (Baltimore: Johns Hopkins University Press, 2016).

5. Delmarva Poultry Industry Inc., "Delmarva's Chicken Business Is Important to You!," www.dpichicken.org/faq_facts/docs/Delmarva%20Chicken%20Production%20Facts%201969-2015.pdf. For Maryland poultry industry facts, see www.dpichicken.org/faq_facts/docs/factsmd2015.pdf. See also Food and Water Watch, "Abusive Poultry Contracts Require Government Action," February 2015, www.foodandwaterwatch.org/sites/default/files/Abusive%20Poultry%20Contracts%20Feb%202015.pdf.

6. Salisbury University Business, Economic and Community Outreach Network Report, "A Scenario Analysis of the Potential Costs of Implementing the Phosphorus Management Tool on the Eastern Shore of Maryland," November 2014, http://mda.maryland.gov/Documents/pmt-analysis.pdf.

7. Environmental Integrity Project, "Poultry's Phosphorus Problem: Phosphorus and Algae in Eastern Shore Waterways: High Concentrations, No Improvement in Past Decade," July 2014, www.environmentalintegrity.org/wp-content/uploads/2016/11/2014-07_Poultrys_Phosphorus_Problem.pdf.

8. Jennifer L. Rhodes, Jennifer Timmons, J. Richard Nottingham, and Wesley Musser, "Broiler Production Management for Potential and Existing Growers," 2011 update, https://extension.umd.edu/sites/default/files/_docs/POULTRY_Broiler ProductionManagement_final1.pdf.

9. Letter from Morison to Holder and Vilsack.

10. Rhodes et al., "Broiler Production Management."

11. Scott Dance, "As Chicken Industry Booms, Eastern Shore Farmers Face Not-in-My-Backyard Activism," *Baltimore Sun*, April 2, 2016, www.baltimoresun.com/news/maryland/bs-md-chicken-farm-growth-20160402-story.html.

12. Food and Water Watch, "Abusive Poultry Contracts."

13. Silbergeld, *Chickenizing Farms and Food.*

14. James M. MacDonald, USDA Economic Research Service, "Technology, Organization, and Financial Performance in U.S. Broiler Production," June 2014, www.ers.usda.gov/publications/pub-details/?pubid=43872.

15. Ibid.

16. Oxfam America, "Lives on the Line: The Human Cost of Cheap Chicken," May 2016, www.oxfamamerica.org/static/media/files/Lives_on_the_Line_Full_Report_Final.pdf.

17. Ibid.

18. Johns Hopkins Bloomberg School of Public Health Center for a Livable Future, videotape of Carole Morison speech at "Public Health in Action" event, April 19, 2016, www.youtube.com/watch?v=G0EW-aiUci8&feature=youtu.be.

19. Perdue press release, "Perdue Expands No Antibiotics Ever Poultry into Mainstream Grocery," 2014, www.perduefarms.com/News_Room/Press_Releases

/details.asp?id=1372&title=Perdue%20Expands%20NO%20ANTIBIOTICS%20EVER%99%20Poultry%20into%20Mainstream%20Grocery.

20. Lance B. Price, Jay P. Graham, Leila G. Lackey, Amira Roess, Rocio Vailes, and Ellen Silbergeld, "Elevated Risk of Carrying Gentamicin-Resistant *Escherichia coli* among U.S. Poultry Workers," *Environmental Health Perspectives* 115, no. 12 (December 2007): 1738–42, www.ncbi.nlm.nih.gov/pmc/articles/PMC2137113/.

21. Justin William Moyer, "Man Arrested after Undercover Video Reveals Alleged Abuse at Perdue Chicken Supplier," *Washington Post*, December 11, 2015, www.washingtonpost.com/news/morning-mix/wp/2015/12/11/man-arrested-after-undercover-video-reveals-alleged-abuse-at-perdue-chicken-supplier/?utm_term=.768941ce7f1b.

22. Perdue Farms Inc. press release, "Perdue Announces Industry-First Animal Care Commitments," June 27, 2016, www.businesswire.com/news/home/201606 27005868/en/ADDING-MULTIMEDIA-Perdue-Announces-Industry-First-Animal-Care. See also Lorraine Mirabella, "Perdue Plans to Reform Its Chicken Welfare Policies," *Baltimore Sun*, June 27, 2016, www.baltimoresun.com/business/bs-bz-perdue-chickens-20160627-story.html.

Ooker Eskridge: Piling Rocks against the Rising Sea

1. Some of the quotes and facts in this chapter are from the *Environment in Focus* radio program "Mayor Fighting Sea-Level Rise Doesn't Believe in Sea-Level Rise," April 27, 2016, http://wypr.org/post/republican-mayor-fighting-sea-level-rise-doesn-t-believe-sea-level-rise#stream/0. Interviews on the island were on April 23 and 24, 2016.

2. Many of the facts about islands in this chapter are from William B. Cronin, *The Disappearing Islands of the Chesapeake* (Baltimore: Johns Hopkins University Press, 2005).

3. Catherine Pierre, "Treasured Islands," *Johns Hopkins Magazine*, November 2005, http://pages.jh.edu/jhumag/1105web/islands.html.

4. Rona Kobell, "Tangier Mayor Hopes That Trump Call Leads to a Seawall," *Bay Journal*, July 4, 2017, www.bayjournal.com/article/trump_weighs_in_on_future_of_tangier.

5. US Geological Survey, "Land Subsidence and Relative Sea-Level Rise in the Southern Chesapeake Bay Region," Circular 1392 (2013), http://pubs.usgs.gov/circ/1392/pdf/circ1392.pdf.

6. Many of the historical facts in this chapter are from exhibits and documents in the Tangier Island Museum. See also Martha W. McCartney, National Park Service, "A Study of Virginia Indians and Jamestown: The First Century," www.nps.gov/parkhistory/online_books/jame1/moretti-langholtz/chap4.htm/; David A. Price, *Love and Hate in Jamestown: John Smith, Pocahontas, and the Start of a New Nation* (New York: Vintage Books, 2003); Bob Deans, *The River Where America Began: A Journey along the James* (Lanham, MD: Rowman & Littlefield, 2007).

THE WILDLIFE
Oysters: Pearl of an Idea: Ban the Oyster Dredge

1. Some of the quotes and facts in this chapter come from three *Environment in Focus* radio programs: (1) "New Oyster Restoration Project Launched in Chesapeake Bay," May 22, 2013, http://programs.wypr.org/podcast/5-22-13-new-oyster -restoration-project-launched-chesapeake-bay; (2) "Multiplying Oyster Harvest Triggers Debate over Catch Restrictions," July 22, 2014, http://wypr.org/post /multiplying-oyster-harvest-triggers-debate-over-catch-restrictions; and (3) "Rise in Power Dredging for Oysters Threatens Fragile Recovery of Bay's Key Species," August 5, 2015, http://wypr.org/post/rise-power-dredging-oysters-threatens -fragile-recovery-bay-s-key-species. The biology of oysters is described in Alice Jane Lippson and Robert L. Lippson, *Life in the Chesapeake Bay: An Illustrated Guide to Fishes, Invertebrates, and Plants of Bays and Inlets from Cape Cod to Cape Hatteras* (Baltimore: Johns Hopkins University Press, 1984). The global picture on oysters is described in a report by Michael W. Beck et al., "Shellfish Reefs at Risk: A Global Analysis of Problems and Solutions," May 2009, www.conservationgateway.org /ConservationPractices/Marine/HabitatProtectionandRestoration/Pages/shell fishreefsatrisk.aspx.

2. Chesapeake Bay Foundation, "On the Brink: Chesapeake's Native Oysters: What It Will Take to Bring Them Back," July 2010, www.cbf.org/document-library /cbf-reports/Oyster_Report_for_Release02a3.pdf.

3. John Wennersten, *The Oyster Wars of Chesapeake Bay* (Washington, DC: Eastern Branch Press, 1981).

4. Chesapeake Bay Foundation, "On the Brink."

5. Interview with Michael Wilberg, professor of biology at the Chesapeake Biological Laboratory of the University of Maryland Center for Environmental Science, April 24, 2014.

6. M. J. Wilberg, M. E. Livings, J. S. Barkman, B. T. Morris, and J. M. Robinson, "Overfishing, Disease, Habitat Loss, and Potential Extirpation of Oysters in Upper Chesapeake Bay," *Marine Ecology Progress Series* 436 (2011): 131–44, http://wilber glab.cbl.umces.edu/pubs/Wilberg%20et%20al%202011.pdf.

7. Maryland Department of Natural Resources press release, "Maryland's Oyster Population Continues to Improve, Highest since 1985," May 7, 2014, http:// news.maryland.gov/dnr/2014/05/07/marylands-oyster-population-continues -to-improve-highest-since-1985/. See also Maryland Department of Natural Resources Oyster Advisory Commission, "Consolidated Strawman Management Plan Proposal," February 13, 2017, http://dnrweb.dnr.state.md.us/fisheries/calendar /events/1259/OAC_Presentation_Consolidated_Proposal.pdf.

8. Interview with Mike Naylor, director of shellfish programs at Maryland Department of Natural Resources, July 21, 2014. See also Maryland Department of Natural Resources, "Maryland Oyster Population Status Report 2015 Fall Survey," http://dnr.maryland.gov/fisheries/Documents/FallSurvey-2015.pdf. In this discus-

sion, the oyster biomass index (on p. 18), an annual estimate of the total weight of oysters in an area, is interpreted as an approximate indicator of oyster populations.

9. Maryland Department of Natural Resources, "Maryland Oyster Population Status Report."

10. Ibid.

11. Quote from *Environment in Focus* radio program "Watermen Evolve into Aquaculture Business Entrepreneurs," November 6, 2013, http://wypr.org/post /watermen-evolve-aquaculture-business-entrepreneurs. Follow-up interviews with the Shockleys via telephone occurred on August 11 and 14, 2015.

12. Maryland Department of Natural Resources, "Oyster Management Review, 2010–2015: Draft Report," July 2010, http://dnr.maryland.gov/fisheries/Docu ments/FiveYearOysterReport.pdf.

13. Tim Wheeler and Rona Kobell, "Watermen Seek, Win, Halt in Tred Avon Oyster Restoration Project," *Bay Journal*, January 13, 2016, www.bayjournal.com /article/watermen_seek_win_halt_in_tred_avon_oyster_restoration_project.

14. Maryland Department of Natural Resources, "DNR Harvest Reserve Areas: Ten Oyster Harvest Reserve Areas Eliminated," January 7, 2016, http:// dnr2.maryland.gov/fisheries/Pages/oysters/index.aspx.

15. Maryland Department of Natural Resources Oyster Advisory Commission, "Consolidated Strawman Management Plan Proposal," February 13, 2017, http://dnrweb.dnr.state.md.us/fisheries/calendar/events/1259/OAC_Presentation _Consolidated_Proposal.pdf.

16. Data on power dredging vs. other methods of harvest received in an e-mail from Frank P. Marenghi, biologist with the Maryland Department of Natural Resources, on August 3, 2015.

17. Bill McKibben, *The End of Nature* (New York: Random House, 1989).

18. Michael J. Wilberg, Maude E. Livings, Jennifer S. Barkman, Brian T. Morris, and Jason M. Robinson, "Overfishing, Disease, Habitat Loss, and Potential Extirpation of Oysters in Upper Chesapeake Bay," *Marine Ecology Progress*, August 31, 2011, www.int-res.com/articles/meps_oa/m436p131.pdf. Quote from Darryl Fears, "Study Calls for Halting Oyster Fishing in Chesapeake Bay," *Washington Post*, September 1, 2011, www.washingtonpost.com/national/health-science/study-calls-for -halting-oyster-fishing-in-chesapeake-bay/2011/09/01/gIQA5EqGvJ_story.html.

19. The Maryland Oyster Advisory Commission's 2008 Report, submitted to the Governor and General Assembly in January 2009, p. 20.

Dermo and MSX: The Parasite Paradox

1. Interview with Mike Naylor, then director of shellfish programs at the Maryland Department of Natural Resources, February 16, 2015.

2. Many of the facts about Dermo and MSX in this chapter are from the Smithsonian Environmental Research Center National Exotic Marine and Estuarine Species Information System, "*Perkinsus marinus*," NEMESIS online database, http://

invasions.si.edu/nemesis/CH-INV.jsp?Species_name=Perkinsus+marinus. See also Chesapeake Bay Foundation, "On the Brink: Chesapeake's Native Oysters: What It Will Take to Bring Them Back," July 2010, www.cbf.org/document-library/cbf -reports/Oyster_Report_for_Release02a3.pdf; Patrick T. K. Woo and Kurt Buchmann, *Fish Parasites: Pathobiology and Protection* (Cambridge, MA: Centre for Agriculture and Bioscience International, 2012).

3. Denise L. Breitburg, Darryl Hondorp, Corinne Audemard, Ryan B. Carnegie, Rebecca B. Burrell, Mark Trice, and Virginia Clark, "Landscape-Level Variation in Disease Susceptibility Related to Shallow-Water Hypoxia," *PLOS One*, February 11, 2015, http://journals.plos.org/plosone/article?id=10.1371/journal.pone.0116223.

4. Some of the quotes and facts in this chapter are from a pair of *Environment in Focus* radio programs: (1) "The Link between 'Dead Zones' and Disease in Oysters," February 25, 2015, http://wypr.org/post/link-between-dead-zones-and-dis ease-oysters; (2) "Multiplying Oyster Harvest Triggers Debate over Catch Restrictions," July 23, 2014, http://wypr.org/post/multiplying-oyster-harvest-triggers -debate-over-catch-restrictions. Interview with Denise Breitburg, senior scientist at the Smithsonian Environmental Research Center, February 13, 2015.

5. Interview with Mike Naylor.

6. Maryland Department of Natural Resources press release, "Maryland's Oyster Population Continues to Improve, Highest since 1985," May 7, 2014, http:// news.maryland.gov/dnr/2014/05/07/marylands-oyster-population-continues-to -improve-highest-since-1985/.

7. Chesapeake Bay Foundation, "On the Brink."

8. Smithsonian Environmental Research Center National Exotic Marine and Estuarine Species Information System, "*Haplosporidium nelsoni,*" NEMESIS online database, http://invasions.si.edu/nemesis/CH-TAX.jsp?Species_name=Haplospori dium%20nelsoni.

Blue Crabs: The Ugly Truth about the Beautiful Swimmers

1. Some of the facts and quotes in this chapter are from four *Environment in Focus* radio programs: (1) "Discovering the Monster Crabs of the Old Chesapeake," January 21, 2015, http://wypr.org/post/discovering-monster-crabs-old-chesapeake; (2) "Cold Weather Deals Blow to Blue Crabs in Bay," August 5, 2014, http://wypr .org/post/cold-weather-deals-blow-blue-crabs-bay-0; (3) "A Perfect Storm for Blue Crabs," April 24, 2012, http://programs.wypr.org/podcast/4-24-12-perfect-storm -blue-crabs; (4) "Red Drum Rise and Blue Crabs Fall," May 29, 2013, http://pro grams.wypr.org/podcast/5/29/13-red-drum-rise-and-blue-crabs-fall.

2. Torben C. Rick, Matthew B. Ogburn, Margaret A. Kramer, Sean T. McCant, Leslie A. Reeder-Myers, Henry M. Miller, and Anson H. Hines, "Archaeology, Taphonomy, and Historical Ecology of Chesapeake Bay Blue Crabs (*Callinectes sapidus*)," *Journal of Archaeological Science* 55 (March 2015): 42–54, www.sciencedirect .com/science/article/pii/S0305440314004774.

3. Tom Horton, *Turning the Tide: Saving the Chesapeake Bay*, 2nd ed. (Washington, DC: Island Press, 2003).

4. Many of the facts about blue crabs in this chapter come from three sources: (1) National Oceanic and Atmospheric Administration Chesapeake Bay Office, "Blue Crab," http://chesapeakebay.noaa.gov/fish-facts/blue-crab; (2) Chesapeake Bay Foundation, *Bad Water and the Decline of Blue Crabs in the Chesapeake Bay* (Chesapeake Bay Foundation, 2008), www.conservationgateway.org/Documents/CBF-BadWatersRe port.pdf; (3) Smithsonian Environmental Research Center, blue crab online resource page, https://serc.si.edu/projects/blue-crab-and-its-fishery.

5. University of Maryland Center for Environmental Science, Chesapeake Bay EcoCheck Report Card for 2011, http://ian.umces.edu/ecocheck/report-cards /chesapeake-bay/2011/.

6. Maryland Department of Natural Resources, "2017 Blue Crab Winter Dredge Survey," http://dnr.maryland.gov/fisheries/Pages/blue-crab/dredge.aspx.

7. Ibid.

8. Ibid.

9. Virginia Marine Resources Commission press release, "Scientific Survey Shows Solid Blue Crab Stock Improvement," April 12, 2016, www.vims.edu/news andevents/topstories/wbcds_2016.php.

10. Interview with Tangier Island waterman Leon McMann, April 23, 2016.

Striped Bass: Recovery and Sickness in Maryland's State Fish

1. Many of the quotes in this chapter were recorded in 2011 and 2012 as part of interviews for the research of three *Environment in Focus* radio programs: (1) "Cry of Loon Fades as Fish Disappear," November 6, 2011, http://programs.wypr.org /podcast/11-16-11-cry-loon-fades-fish-disappear; (2) "Time Running Out for Research into Bay Disease," April 18, 2012, http://programs.wypr.org/category/pod cast-keywords/mycobacteriosis; (3) "The Recovery and Decline of Maryland's State Fish," May 4, 2011, http://programs.wypr.org/podcast/5-4-11-recovery-and-decline -maryland%E2%80%99s-state-fish.

2. Dick Russell, *Striper Wars: An American Fish Story* (Washington, DC: Island Press, 2006).

3. Many of the facts about striped bass in this chapter are from two sources: (1) Chesapeake Bay Program, "Striped Bass," www.chesapeakebay.net/fieldguide /critter/striped_bass; (2) National Oceanic and Atmospheric Administration Chesapeake Bay, "Striped Bass," http://chesapeakebay.noaa.gov/fish-facts/striped-bass.

4. Russell, *Striper Wars*.

5. Chesapeake Bay Program, "Striped Bass Abundance," www.chesapeakebay .net/indicators/indicator/striped_bass_abundance.

6. A list of scientific journal articles by Wolfgang Vogelbein and colleagues on mycobacteriosis can be found on the Virginia Institute of Marine Science website; see www.vims.edu/research/departments/eaah/programs/projects/myco/index.php.

7. E-mail communication from Wolfgang Vogelbein of the Virginia Institute of Marine Science, February 24, 2016. His federal funding ran out in 2012.

American Eels: River of Trouble for a Backward Fish

1. Some of the quotes and details in this chapter come from three sources: (1) Tom Pelton, "Chesapeake Eels: To Some, Bait; to Others, an Exportable Delicacy," *Baltimore Sun*, April 8, 2006, http://articles.baltimoresun.com/2006-04-08/news/0604080250_1_american-eel-benedict-chesapeake (with permission for reprinted portions); (2) *Environment in Focus* radio program "The Eelman," April 30, 2008; (3) *Environment in Focus* radio program "Stream of Troubles for a Backward Fish," June 1, 2011.

2. Atlantic States Marine Fisheries Commission, "American Eel," www.asmfc.org/species/american-eel.

3. Ibid.

4. Maryland Department of Natural Resources Fisheries Service, "2015 Fishery Management Plans: Report to the Legislative Committees," December 2016, http://dnr.maryland.gov/fisheries/Documents/Full_FMP_2016.pdf.

5. E-mail from David Secor, University of Maryland Center for Environmental Science Chesapeake Biological Lab biologist, May 26, 2016.

6. E-mail from Steve Minkkinen, biologist with the US Fish and Wildlife Service, May 26, 2016.

Sturgeon: The Dinosaur Matchmaker Meets His Match

1. Mike Mangold, Sheila Eyler, Steve Minkkinen, and Brian Richardson, US Fish and Wildlife Service and Maryland Department of Natural Resources, "Atlantic Sturgeon Reward Program for Maryland Waters of the Chesapeake Bay and Tributaries, 1996–2006," November 2007, www.fws.gov/northeast/marylandfisheries/reports/REWARDPROGRAMPAPERFINAL.pdf.

2. The events at the opening of this story happened in 2005, and Lazur later left the University of Maryland Center for Environmental Science Horn Point Lab in 2010 to become director of Maryland Sea Grant at the University of Maryland. Some of the quotes and details in this chapter are from a pair of *Baltimore Sun* articles and an *Environment in Focus* radio program on the subject of sturgeon: (1) Tom Pelton, "Sturgeon Showing Its Survival Skills," *Baltimore Sun*, May 24, 2005, http://articles.baltimoresun.com/2005-05-24/news/0505240161_1_sturgeon-chesapeake-bay-fish; (2) Tom Pelton, "Rare Catch Could Spawn Resurgence of Sturgeon in Bay," *Baltimore Sun*, May 7, 2007, http://articles.baltimoresun.com/2007-06-13/news/0706130110_1_sturgeon-chesapeake-bay-university-of-maryland; (3) "The Sturgeon Matchmaker," January 23, 2008.

3. Many of the historical facts in this chapter are from Inga Saffron, *Caviar: The Strange History and Uncertain Future of the World's Most Coveted Delicacy* (New York: Broadway, 2003).

4. National Oceanic and Atmospheric Administration Fisheries press release, "NOAA Lists Five Atlantic Sturgeon Populations under Endangered Species Act," January 31, 2012, www.nmfs.noaa.gov/stories/2012/01/31_atlantic_sturgeon.html.

5. Rachel Uda, "Atlantic Sturgeon Making Hudson River Comeback, DEC Says," *New York Newsday*, February 3, 2016, www.newsday.com/news/new-york/atlantic-sturgeon-making-hudson-river-comeback-dec-says-1.11425808.

6. New York Department of Environmental Conservation press release, "DEC: Atlantic Sturgeon Show Encouraging Signs for Population Recovery," February 3, 2016, http://www.dec.ny.gov/press/104944.html.

7. Maryland Department of Natural Resources press release, "Mature Endangered Atlantic Sturgeon Discovered in Marshyhope Creek," September 17, 2014, http://news.maryland.gov/dnr/2014/09/17/mature-endangered-atlantic-sturgeon-discovered-in-marshyhope-creek/.

8. Some of the quotes and observations from this part of the chapter are from the *Environment in Focus* radio program "The Fish That Saved Jamestown," November 24, 2010.

9. Virginia Commonwealth University, "Atlantic Sturgeon Restoration," www.ricerivers.vcu.edu/research/atlantic-sturgeon-restoration/.

10. Melissa Lesh, Virginia Commonwealth University, online video on James River Sturgeon Project, https://vimeo.com/81156656.

THE POLICIES
Enforcement: Words versus Water

1. United States of America and State of Maryland v. Mayor and City Council of Baltimore, Civil Action No. JFM-02-1524, Consent Decree, www.baltimorecity.gov/sites/default/files/Consent%20Decree.pdf.

2. City of Baltimore Wastewater Utility Fund annual financial statements, 2002 to 2014. Available by request from the Baltimore Comptroller's Office.

3. Baltimore sewage overflow figures are from two sources: (1) the Maryland Department of the Environment online database of reported sewage overflows and spills, which is available at http://mde.maryland.gov/programs/water/Compliance/Pages/ReportedSewerOverflow.aspx; and (2) spreadsheets from the Baltimore Department of Public Works (DPW) for metered discharges from outfalls SSO #67 and SSO #72 on the Jones Falls, which were provided to the author in response to a Maryland Public Information Act request. A summary of the state and city data can be found in the Environmental Integrity Project report "Stopping the Flood beneath Baltimore's Streets," funded by the Abell Foundation, which I coauthored, released December 15, 2015; see http://environmentalintegrity.org/wp-content/uploads/FINAL-SEWAGE-REPORT.pdf. The city's progress report to MDE and EPA for January 2016 is available online at http://publicworks.baltimorecity.gov/sites/default/files/Quarterly%20Report%2053%20full.pdf. Briefing on an update to decree progress was provided verbally by DPW officials during a public

meeting on June 7, 2016, at Maryland Department of the Environment headquarters.

4. Modified Baltimore sewage consent decree. United States of America and
the State of Maryland v. Mayor and City Council of Baltimore, Civil Action No.
JFM-02-1524, www.justice.gov/sites/default/files/enrd/pages/attachments/2016
/06/01/filed_baltimore_modifed_consent_decree.pdf.

5. Baltimore Waterfront Partnership, "Swimmable and Fishable by 2020,"
http://baltimorewaterfront.com/healthy-harbor/.

6. Environmental Integrity Project, "Stopping the Flood."

7. Information from the *Environment in Focus* radio program "Baltimore Releases 15 X More Sewage Than Reported to Public," December 16, 2015, http://
wypr.org/post/baltimore-releases-15-x-more-sewage-reported-public.

8. EPA press release, "U.S., Maryland Amend Agreement with Baltimore City
to Curtail Sewer Overflows and Improve Water Quality Requires Annual Public
Progress Reporting," June 1, 2016, www.epa.gov/newsreleases/us-maryland-amend
-agreement-baltimore-city-curtail-sewer-overflows-and-improve-water.

9. Environmental Integrity Project, "Stopping the Flood."

10. Ibid. Data on bacteria levels are from the Baltimore Department of Public
Works Stream Impact Sampling program.

11. Environmental Integrity Project, "Stopping the Flood."

12. Ibid.

13. Ibid.

14. Ibid.

15. Ibid.

16. Ibid.

17. Ibid. Some of the quotes and details in this chapter also appeared in the
Environment in Focus radio program "Baltimore Releases 15 X More Sewage."

18. Environmental Integrity Project, "Stopping the Flood."

19. Ibid.

20. Ibid.

Pennsylvania: The Bay Cleanup Collides with the Politics of Rural America

1. US Environmental Protection Agency, "Chesapeake Bay Total Maximum
Daily Load (TMDL) for Nitrogen, Phosphorous and Sediment," www.epa.gov/sites
/production/files/2014-12/documents/cbay_final_tmdl_exec_sum_section_1
_through_3_final_0.pdf.

2. TMDL Section 4, www.epa.gov/sites/production/files/2014-12/documents
/cbay_final_tmdl_section_4_final_0.pdf.

3. Pennsylvania Department of Environmental Protection, "A DEP Strategy to
Enhance Pennsylvania's Bay Restoration Effort," January 21, 2016, www.dep.state
.pa.us/river/iwo/chesbay/docs/DEPChesapeakeBayRestorationStrategy012116.pdf.

4. TMDL (see note 1 above) and also Pennsylvania Department of Environmental Protection, "DEP Strategy."

5. Pennsylvania Department of Environmental Protection, "DEP Strategy."

6. "Average-sized" for Pennsylvania in this context means about 75 to 275 cows and about 130 acres of crops to feed the animals.

7. Tom Horton, *Turning the Tide: Saving the Chesapeake Bay*, 2nd ed. (Washington, DC: Island Press, 2003).

8. US Environmental Protection Agency, "EPA Evaluation of Pennsylvania's 2014–2015 and 2016–2017 Milestones," June 2016, www.epa.gov/sites/production /files/2016-06/documents/pa_2014-2015_-_2016-2017_milestone_eval_06-17 -16.pdf.

9. Chesapeake Bay Foundation, "Angling for Healthier Rivers," 2013.

10. US Geological Survey, "Summary of Trends Measured at the Chesapeake Bay Tributary Sites: Water Year 2013 Update," December 5, 2014.

11. September 14, 2015, interview of Jeff Corbin, then the Chesapeake Bay advisor to the EPA administrator, for the *Environment in Focus* radio program "EPA at Crossroads over Pennsylvania's Failure to Control Farm Pollution," September 16, 2015, http://wypr.org/post/epa-crossroads-over-pennsylvania-s-failure-control -farm-pollution. Corbin has since changed jobs.

12. Rona Kobell, "EPA Chief Calls Pennsylvania's Lagging Bay Cleanup 'discouraging,'" *Bay Journal*, May 25, 2016, www.bayjournal.com/blog/post/epa_chief _call_pennsylvanias_lagging_bay_cleanup_discouraging.

13. US Environmental Protection Agency, "EPA Evaluation of Pennsylvania's 2014–2015 and 2016–2017 Milestones."

14. Pennsylvania upgraded its sewage plants to so-called Biological Nutrient Removal (BNR) level instead of Enhanced Nutrient Removal (ENR).

15. US Environmental Protection Agency, "EPA Chesapeake Bay Total Maximum Daily Load."

16. Numbers as of May 1, 2015. "State-of-the-art" is defined as ENR systems, and "highest water quality standards" is defined here as less than 5 milligrams per liter of dissolved nitrogen. Pennsylvania has upgraded many of its sewage plants to a lower level, known as BNR. Data on wastewater treatment plant upgrades were provided by the EPA Chesapeake Bay Program via e-mail on May 1, 2015.

17. Maryland Department of the Environment, "Bay Restoration Fund Targeted Wastewater Treatment Plants, Reporting Period: May 2017," www.mde .state.md.us/programs/water/BayRestorationFund/Documents/5-BRF-WWTP %20Update%20for%20BayStat.pdf.

18. Chesapeake Bay Commission, "Healthy Livestock, Healthy Streams: Policy Actions to Promote Livestock Stream Exclusion," May 2015, www.chesbay.us/Pub lications/Healthy%20Livestock,%20Healthy%20Streams.pdf.

19. Pennsylvania DEP press release, "Wolf Administration Announces Successful First Year for Expanded Agricultural Inspections in Chesapeake Bay Watershed,"

August 16, 2017, http://www.ahs.dep.pa.gov/NewsRoomPublic/articleviewer.aspx? id=21272&typeid=1. See also Karl Blankenship, "PA Plan Says It Will Increase Ag Inspections, Plant More Trees," *Bay Journal*, February 28, 2016, www.bay journal.com/article/pa_plan_says_it_will_increase_ag_inspections_plant_more _trees.

20. Daniel Walmer, "DEP Targets Farmers to Reduce Pollution," *Lebanon Daily News*, May 11, 2016, www.ldnews.com/story/news/local/2016/05/11/dep -targets-farmers-reduce-pollution-chesapeake-bay/84189268/.

21. Maryland Department of Agriculture, "Fiscal Year 2015 Annual Report, MDA Nutrient Management Program," March 16, 2015, http://mda.maryland.gov /resource_conservation/counties/NMPAnnualReport%202015FINAL_web.pdf.

22. Michelle R. Perez, "Regulating Farmer Nutrient Management: A Three-State Case Study on the Delmarva Peninsula," *Journal of Environmental Quality* 44, no. 2 (March 2015): 402–14, https://dl.sciencesocieties.org/publications/jeq/pdfs /44/2/402.

23. Maryland Department of Agriculture, "Fiscal Year 2015 Annual Report."

24. Environmental Integrity Project report, "Manure Overload on Maryland's Eastern Shore," December 8, 2014, www.environmentalintegrity.org/wp-content /uploads/2016/11/Manure-Overload1.pdf. See also US Geological Survey, "Understanding Nutrients in the Chesapeake Bay Watershed and Implications for Management and Restoration—the Eastern Shore," March 2015, https://pubs.usgs .gov/circ/1406/pdf/circ1406.pdf.

25. Michael J. Langland, "Sediment Transport and Capacity Change in Three Reservoirs, Lower Susquehanna River Basin, Pennsylvania and Maryland, 1900–2012," US Geological Survey, 2014, http://dx.doi.org/10.3133/ofr20141235. Calculated using the estimate that the reservoir is 92% full of silt. See also USGS press release, "Conowingo Dam above 90 Percent Capacity for Sediment Storage," February 18, 2015, www.usgs.gov/news/conowingo-dam-above-90-percent-capacity -sediment-storage.

26. Maryland Sea Grant, "A 2011 Storm Walloped the Bay with Sediment," July 30, 2013, www.mdsg.umd.edu/news/2011-storm-walloped-bay-sediment-study-says.

27. Robert M. Hirsch, "Flux of Nitrogen, Phosphorus, and Suspended Sediment from the Susquehanna River Basin to the Chesapeake Bay during Tropical Storm Lee, September 2011, as an Indicator of the Effects of Reservoir Sedimentation on Water Quality," US Geological Survey, 2012, http://pubs.usgs.gov/sir /2012/5185/pdf/sir2012-5185-508.pdf.

28. Pennsylvania Department of Environmental Protection, "The Pennsylvania Chesapeake Bay Strategy: Improving Local Water Quality in Pennsylvania and Restoring the Chesapeake Bay," January 21, 2016, www.dep.pa.gov/Business/Water /Pages/Chesapeake-Bay-Office.aspx#.VwG3KXjD_IU.

29. Telephone interview with John Quigley, secretary of the Pennsylvania Department of Environmental Protection, April 1, 2016.

30. Telephone interview with Mark O'Neill, spokesman for the Pennsylvania Farm Bureau, April 4, 2016. This and other quotes and observations in this chapter also appeared in the *Environment in Focus* radio program "70 Percent of Pennsylvania Farmers Violate Clean Water Law," April 6, 2016, http://wypr.org/post/70 -percent-pennsylvania-farmers-violate-clean-water-law.

31. Associated Press, "What Will History Say of Gov. Tom Corbett's Tenure in Pennsylvania?," December 25, 2014, www.pennlive.com/politics/index.ssf/2014 /12/what_will_history_say_of_gov_t.html.

32. Telephone interview with John Quigley.

33. Telephone interview with Evan Isaacson, policy analyst for the Center for Progressive Reform, April 4, 2016.

34. Telephone interview with Jacqueline Bonomo, chief operating officer of Penn Future, April 4, 2016.

35. Telephone interview with EPA Region 3 water program director Jon Capacasa, September 16, 2016, conducted for the *Environment in Focus* radio program.

36. E-mail from David Sternberg, spokesman for EPA Region 3 office in Philadelphia, September 19, 2016.

37. Ibid.

38. Interview with Jeff Corbin, who was then senior advisor to the EPA administrator for the Chesapeake Bay and Anacostia River (the EPA bay "czar"), September 14, 2015.

Air Pollution versus Water Pollution: Cleaning the Water from the Sky

1. Many of the descriptions and quotes in this chapter are from the *Environment in Focus* radio program "Cleaning Up the Water from the Sky," October 5, 2016, http://wypr.org/post/cleaning-water-sky.

2. Keith N. Eshleman and Robert D. Sabo, "Declining Nitrate-N Yields in the Upper Potomac River Basin: What Is Really Driving Progress under the Chesapeake Bay Restoration?," *Atmospheric Environment* 146 (2016): 280–89, www.umces .edu/sites/default/files/Eshleman%20study_Atmospheric%20Environment.pdf.

3. US Environmental Protection Agency, "Benefits and Costs of the Clean Air Act 1990–2020, the Second Prospective Study," March 2011, www.epa.gov/clean -air-act-overview/benefits-and-costs-clean-air-act-1990-2020-second-prospective -study.

4. Chesapeake Bay Foundation, "Debunking the 'Job Killer' Myth: How Pollution Limits Encourage Jobs in the Chesapeake Bay Region," December 2011, www .cbf.org/document-library/cbf-reports/Jobs-Report-120103-FINALe2ef.pdf.

5. Ibid.

6. Ibid.

7. US Environmental Protection Agency, "Air Pollution in the Chesapeake Bay Watershed," www.epa.gov/chesapeake-bay-tmdl/air-pollution-chesapeake-bay-watershed.

8. US Geological Survey and partners, "New Insights: Science-Based Evidence of Water Quality Improvements, Challenges, and Opportunities in the Chesapeake," March 2014, www.chesapeakebay.net/channel_files/21409/new_insights_report.pdf.

9. US Environmental Protection Agency, "Assessment of the Potential Impacts of Hydraulic Fracturing for Oil and Gas on Drinking Water Resources," June 2015, www.epa.gov/sites/production/files/2015-06/documents/hf_es_erd_jun2015.pdf. See also EPA Science Advisory Board, "SAB Review of the EPA's Draft Assessment of the Potential Impacts of Hydraulic Fracturing for Oil and Gas on Drinking Water Resources," February 16, 2016, https://yosemite.epa.gov/sab/sabproduct.nsf/4CF68F09429EDF0C85257F5B0061DFDB/$File/2-16-16+Draft+SAB+HF+Rpt+(Redline).pdf; Garth T. Llewellyn, Frank Dorman, J. L. Westland, D. Yoxtheimer, Paul Grieve, Todd Sowers, E. Humston-Fulmer, and Susan L. Brantley, "Evaluating a Groundwater Supply Contamination Incident Attributed to Marcellus Shale Gas Development," *Proceedings of the National Academy of Sciences* 112, no. 20 (May 2015): 6325–30, www.pnas.org/content/112/20/6325.

10. Telephone interview with William Dennison, vice president of UMCES, May 19, 2016.

Agriculture: A Tale of Two Farmers

1. Some of the quotes, observations, and facts in this chapter come from a pair of *Environment in Focus* radio programs: (1) "Farmers Buck Chesapeake Cleanup Rules," July 20, 2016, http://wypr.org/post/farmers-buck-chesapeake-cleanup-rules; (2) "Picking Food That Is Healthy for the Chesapeake Bay," June 29, 2016, http://wypr.org/post/picking-food-healthy-chesapeake-bay.

2. US Environmental Protection Agency Chesapeake Bay Program, "Agriculture," www.chesapeakebay.net/issues/issue/agriculture#inline.

3. Environmental Working Group Farm Subsidy database, https://farm.ewg.org/region.php?fips=24000.

4. Ibid.

5. EPA, "Guidance for Federal Land Management in the Chesapeake Bay Watershed," May 12, 2010, www.epa.gov/sites/production/files/2015-10/documents/chesbay_chap01.pdf.

6. US Environmental Protection Agency Chesapeake Bay Program, "Agriculture."

7. Karl Blankenship and Whitney Pipkin, "Census: Farmland Growing in Bay States: Increase in Acreage Has Implications for Bay Restoration Strategies," *Bay Journal*, July 20, 2014, www.bayjournal.com/article/census_farmland_growing_in_bay_states.

8. Mike Lavender, "Report Highlights Corn Ethanol's Devastating Toll," *Environmental Working Group*, September 5, 2014, www.ewg.org/agmag/2014/09/report-highlights-corn-ethanol-s-devastating-toll.

9. Delmarva Poultry Industry Inc., "Facts about Delmarva's Meat Chicken Industry," www.dpichicken.org/faq_facts/docs/Delmarva%20Chicken%20Production%20Facts%201969-2015.pdf.

10. US Department of Agriculture, "2012 Census of Agriculture," www.agcensus.usda.gov/Publications/2012/Full_Report/Volume_1,_Chapter_1_State_Level/Maryland/st24_1_004_005.pdf. Note: farm income can be difficult to compare to nonfarm income, because of differences in deductions allowed for each.

11. Maryland Dairy Advisory Council, "Maryland's Dairy Industry: 2012," November 2012, http://mda.maryland.gov/Documents/2012_Dairy_Council_Report.pdf.

12. Interview with Chuck Fry, president of the Maryland Farm Bureau, July 15, 2016.

13. Quote from the *Environment in Focus* radio program "The War on Rural Maryland," February 22, 2012, http://programs.wypr.org/podcast/2-22-12-war-rural-maryland.

14. Ike Wilson, "Tuscarora Dairy Farmer Selected to Lead Maryland Farm Bureau," *Frederick News Post*, December 17, 2013, http://bit.ly/2aD3ZO4; John R. Wennersten, *The Chesapeake: An Environmental Biography* (Baltimore: Maryland Historical Society Press, 2001).

15. Chuck Fry, president of the Maryland Farm Bureau, estimated that perhaps half of farmers with cattle fence them out of streams; interview with Chuck Fry, July 15, 2016. An estimate of 27% comes from David Fahrenthold, "Scenes of an Effort Impeded Unfold across Chesapeake Watershed," *Washington Post*, December 26, 2008, www.washingtonpost.com/wp-dyn/content/article/2008/12/26/AR2008122601710.html.

16. Numbers provided by Shenandoah Riverkeeper Mark Frondorf via e-mail on April 10, 2017.

17. Virginia Department of Conservation and Recreation, "The Bottom Line: It Pays to Exclude Livestock from Streams," www.dcr.virginia.gov/soil-and-water/document/cmw/fsstreamexcl.pdf.

18. Chesapeake Bay Commission, "Healthy Livestock, Healthy Streams: Policy Actions to Promote Livestock Stream Exclusion," May 2015, www.chesbay.us/Publications/Healthy%20Livestock,%20Healthy%20Streams.pdf.

19. Maryland Department of Agriculture, "Fiscal Year 2015 Annual Report, Nutrient Management Program," March 16, 2016, http://mda.maryland.gov/resource_conservation/counties/NMPAnnualReport%202015FINAL_web.pdf; US Environmental Protection Agency Region 3 report, "Pennsylvania Animal Agriculture Program Assessment," February 2015, www.epa.gov/sites/production/files/2015-07/documents/pennsylvania_animal_agriculture_program_assessment_final_2.pdf. See also Karl Blankenship, "PA Plan Says It Will Increase Ag Inspections, Plant More Trees," *Bay Journal*, February 28, 2016, www.bayjournal.com/article/pa_plan_says_it_will_increase_ag_inspections_plant_more_trees.

20. Michelle R. Perez, "Regulating Farmer Nutrient Management: A Three-State Case Study on the Delmarva Peninsula," *Journal of Environmental Quality* 44, no. 2 (March 2015): 402–14, https://dl.sciencesocieties.org/publications/jeq/pdfs /44/2/402.

21. Among the studies that link neonicotinoids to bee declines is Ignaz Wessler, Hedwig-Annabel Gärtner, Rosmarie Michel-Schmidt, Christoph Brochhausen, Luise Schmitz, Laura Anspach, Bernd Grünewald, and Charles James Kirkpatrick, "Honeybees Produce Millimolar Concentrations of Non-neuronal Acetylcholine for Breeding: Possible Adverse Effects of Neonicotinoids," *PLOS One*, June 10, 2016, http://journals.plos.org/plosone/article?id=10.1371/journal.pone.0156886.

22. Kathryn Z. Guyton, Dana Loomis, Yann Grosse, Fatiha El Ghissassi, Lamia Benbrahim-Tallaa, Neela Guha, Chiara Scoccianti, Heidi Mattock, and Kurt Straif, on behalf of the International Agency for Research on Cancer Monograph Working Group, IARC, Lyon, France, "Carcinogenicity of Tetrachlorvinphos, Parathion, Malathion, Diazinon, and Glyphosate," *Lancet* 16, no. 5 (March 2015): 490–91, www .thelancet.com/journals/lanonc/article/PIIS1470-2045(15)70134-8/abstract.

23. Brice X. Semmens, Darius J. Semmens, Wayne E. Thogmartin, Ruscena Wiederholt, Laura López-Hoffman, Jay E. Diffendorfer, John M. Pleasants, Karen S. Oberhauser, and Orley R. Taylor, "Quasi-extinction Risk and Population Targets for the Eastern, Migratory Population of Monarch Butterflies (*Danaus plexippus*)," Nature Scientific Reports 6, Article no. 23265, March 21, 2016, www.nature.com /articles/srep23265. See also Sarah P. Saunders, Leslie Ries, Karen S. Oberhauser, Wayne E. Thogmartin, and Elise F. Zipkin, "Local and Cross-Seasonal Associations of Climate and Land Use with Abundance of Monarch Butterflies *Danaus plexippus*," *Ecography*, May 16, 2017, http://onlinelibrary.wiley.com/doi/10.1111/ecog .02719/abstract;jsessionid=BC32FABFB7A3ADEF56E0FA3DF26C265A.f02t03.

24. World Health Organization, "Obesity and Overweight Fact Sheet," updated June 2016, www.who.int/mediacentre/factsheets/fs311/en/.

25. Chesapeake Bay Program, "Planting Forested Buffers," www.chesapeakebay .net/indicators/indicator/planting_forest_buffers.

26. Interview with Sally Claggett, US Forest Service liaison with the Chesapeake Bay Program, June 24, 2016.

27. Ibid.

28. Interview with Tom Simpson, senior scientist at Aqua Terra Science LLC, an environmental consulting firm, July 15, 2016.

Climate Change: At War with a Changing Climate

1. Some of the quotes and observations in this chapter are from the *Environment in Focus* radio program "At War with a Changing Climate," October 17, 2016, http://wypr.org/post/war-changing-climate.

2. Forbes Tompkins and Christina DeConcini of the World Resources Institute, "Sea-Level Rise and Its Impact on Virginia," September 2016, www.wri.org/sites /default/files/wri_factsheet_virginia_final.pdf.

3. Interview with Michael King, Regional Community Planning liaison officer, Navy Region Mid-Atlantic, September 31, 2016.

4. Tompkins and DeConcini, "Sea-Level Rise."

5. Maryland Commission on Climate Change, Scientific and Technical Working Group report, "Global Warming and the Free State: Comprehensive Assessment of Climate Change Impacts in Maryland," 2008, https://pdfs.semanticscholar.org /83ac/548ae5e98b691a5194a92056137e6eb3911d.pdf.

6. Ibid.

7. Ibid.

8. US Fish and Wildlife Service, "Blackwater National Wildlife Refuge Marsh Loss and Restoration," 2009, www.fws.gov/northeast/climatechange/pdf/black watermarshloss122009.pdf.

9. US Geological Survey, "Land Subsidence and Relative Sea-Level Rise in the Southern Chesapeake Bay Region," 2013, http://pubs.usgs.gov/circ/1392/pdf /circ1392.pdf.

10. Ibid.

Advocacy and Pollution Trading: How "Save the Bay" Became "Trade the Bay"

1. Chesapeake Bay Foundation, "Our History," www.cbf.org/about-cbf/history.

2. Ibid.

3. At the time this book was written, in 2017, Maryland governor Larry Hogan, a Republican, had a mixed environmental record and so the jury was still out on him. While his administration approved new regulations for the management of poultry waste, it also weakened rules designed to reduce runoff of manure from farm fields. Governor Hogan signed a bill to ban hydraulic fracturing, but the bill had been sponsored and advanced by Democratic lawmakers and Hogan only endorsed it after enough Democrats backed it to create a veto-proof majority. Hogan also successfully pushed for an elimination of Maryland's "rain tax" stormwater fee mandate for local governments, which was a step backward for the bay cleanup.

4. On March 3, 2016, presidential candidate Donald Trump pledged during a Republican primary debate in Detroit, "Department of Environmental Protection: We are going to get rid of it in almost every form." See www.youtube.com /watch?v=EOptTq6WYKY. See also Chris Kaufman, "Republican Trump Says 70 Percent of Federal Regulations 'Can Go,'" *Reuters*, October 7, 2016, www.reuters .com/article/us-usa-election-trump-regulations-idUSKCN12629R; Louis Jacobson, "Yes, Donald Trump Did Call Climate Change a Chinese Hoax," *Politifact*, June 3, 2016, www.politifact.com/truth-o-meter/statements/2016/jun/03/hillary-clin ton/yes-donald-trump-did-call-climate-change-chinese-h/.

5. Juliet Eilperin and Brady Dennis, "White House Eyes Plan to Cut EPA Staff by One-Fifth, Eliminating Key Programs," *Washington Post*, March 1, 2017, www .washingtonpost.com/news/energy-environment/wp/2017/03/01/white-house

-proposes-cutting-epa-staff-by-one-fifth-eliminating-key-programs/?utm_term=
.d5c5fcb844aa.

6. Karl Blankenship, "House Votes to Restrict EPA Oversight Power in Bay Cleanup," *Bay Journal*, July 14, 2016, www.bayjournal.com/blog/post/house_votes _to_restrict_epa_oversight_power_in_bay_cleanup.

7. CBF annual reports available at www.cbf.org/about-cbf/financials/index .html.

8. Chesapeake Bay Foundation Statement of Financial Position, as of June 30, 2016, www.cbf.org/document-library/financial-documents/2016-audited-financial .pdf.

9. This theme is explored in Howard R. Ernst, *Chesapeake Bay Blues: Science, Politics, and the Struggle to Save the Bay* (New York: Rowman & Littlefield, 2003), and *Fight for the Bay: Why a Dark Green Environmental Awakening Is Needed to Save the Chesapeake Bay* (New York: Rowman & Littlefield, 2010).

10. Governor Martin O'Malley, a Democrat, promised when he was running for office to "fully fund" a successful Maryland land preservation program called Program Open Space. But then O'Malley turned out to be similar to his Republican predecessor, Governor Robert Ehrlich, in transferring real estate transfer tax money out of what was supposed to be a dedicated fund just for protecting fields and forests and building playgrounds. To help fund the general operations of government, O'Malley transferred $172 million out of Program Open Space in 2009 and $144 million in 2013 and 2014, according to the Maryland Department of Legislative Services. O'Malley replaced some of these funds with borrowed money, but like Ehrlich, he undermined a highly effective land conservation law. In 2007, O'Malley and the Maryland General Assembly created the Chesapeake and Atlantic Coastal Bays Trust Fund to pay for projects to reduce runoff pollution. But then in 2011, O'Malley transferred $60 million in rental car and gas taxes out of what was supposed to be a fund dedicated just for bay cleanup. For details, read the Maryland Department of Legislative Services report, "The 90 Day Report: A Review of the 2011 Legislative Session," http://mgaleg.maryland.gov/Pubs/legislegal/2011rs -90-day-report.pdf.

11. Oyster Recovery Partnership, "Board and Staff," 2016, http://oysterrecovery .org/boardstaff/.

12. Chesapeake Bay Foundation, "Decades of Success," www.cbf.org/decades-of -success/2010s.

13. Tom Pelton, "Bay Foundation Plans to Support Farmers," *Baltimore Sun*, September 21, 2005, http://articles.baltimoresun.com/2005-09-21/news/0509 210143_1_chesapeake-bay-foundation-maryland-farmers-maryland-farm-bureau.

14. The total includes both US Department of Agriculture funds and federal economic "stimulus" funds from the Obama administration, much of which CBF then passed on to contractors and farmers. In 2009, the federal government (through a Pennsylvania agency called PENNVEST) gave CBF $14.9 million in American Recovery and Reinvestment Act funds to pay local contractors to install a variety

of farm conservation practices that will reduce pollution and, in some cases, use manure to create energy. In 2016, CBF and its partners received $1.1 million from the US Department of Agriculture (USDA) to help livestock farmers in Maryland carry out conservation practices like raising animals on pasture instead of in confined areas. In 2015, the USDA's Natural Resources Conservation Service (NRCS) gave CBF $491,070 to expand the use of management-intensive grazing in the Chesapeake Bay watershed, including enrolling at least 35 farmers to transition 1,400 acres of farmland to rotation grazing. In 2012, the NRCS gave CBF a $700,880 grant for "operationalizing water quality trading in the Chesapeake Bay." In 2011, the NRCS gave CBF a $454,797 grant "to develop and implement a greenhouse gas tool for estimating N_2O reductions from nutrient management in the Chesapeake Bay watershed. In 2009, the NRCS gave CBF $800,314 for "Innovative Ways to Increase Ag BMP Adoption." In 2008, the NRCS gave CBF a $500,000 grant to support the "The Chesapeake Nutrient Neutral Fund," including for "nutrient credit standards & registry." In 2005, the NRCS gave CBF $400,298 for "Precision Dairy Feeding to Reduce Nutrient Pollution in Pennsylvania's Waters and the Chesapeake Bay."

15. Chesapeake Bay Foundation, "Decades of Success."

16. When I worked as senior writer and investigative reporter at CBF, the organization declined to release a report I researched and wrote about agricultural chemical fertilizer pollution in the bay because top managers said they did not want to anger the farm lobby or the Koch brothers. Charles and David Koch are major Republican donors who also own a fertilizer manufacturing company, Koch Fertilizer LLC. CBF also refused to consider report proposals about the poultry industry, even though it is a major source of pollution in the bay.

17. Rona Kobell, "Bay Advocates Called Soft on Farm Pollution," *Baltimore Sun*, June 9, 2008, http://articles.baltimoresun.com/2008-06-09/news/0806080254_1 _chesapeake-bay-foundation-pollution-poultry-farmers.

18. Chesapeake Bay Foundation, "Senate Bill 1029—the Maryland Agricultural Certainty Program—Passes," 2013, www.legacy-cbf.org/about-cbf/offices -operations/annapolis-md/the-issues/sb-1029-agricultural-certainty-30/15/13.

19. Karl Blankenship, "EPA, CBF Agree to Not Seek New National CAFO Regulations," *Bay Journal*, July 4, 2013, www.bayjournal.com/article/epa_cbf_agree _to_not_seek_new_national_cafo_regulations.

20. Chesapeake Bay Foundation press release, "$28 Million in New Federal and State Funding Can Jumpstart Pennsylvania's Clean Water Efforts," October 4, 2016, www.cbf.org/news-media/newsroom/pa/2016/10/04/28-million-in-new -federal-and-state-funding-can-jumpstart-pennsylvanias-clean-water-efforts.

21. Karl Blankenship, "PA Plan Says It Will Increase Ag Inspections, Plant More Trees," *Bay Journal*, February 28, 2016, www.bayjournal.com/article/pa _plan_says_it_will_increase_ag_inspections_plant_more_trees. See also Pennsylvania DEP press release, "Wolf Administration Announces Successful First Year for Expanded Agricultural Inspections in Chesapeake Bay Watershed," August 16,

2017, http://www.ahs.dep.pa.gov/NewsRoomPublic/articleviewer.aspx?id=21272 &typeid=1.

22. Chesapeake Bay Foundation grant application, "Operationalizing Water Quality Trading in the Chesapeake Bay: A Proposal to the USDA Natural Resources Conservation Service (NRCS) by the Chesapeake Bay Foundation (CBF) Conservation Innovative Grant (CIG)—Chesapeake Bay Water Quality Trading," March 2, 2012.

23. Richard Conniff, "The Political History of Cap and Trade: How an Unlikely Mix of Environmentalists and Free-Market Conservatives Hammered Out the Strategy Known as Cap-and-Trade," *Smithsonian*, August 2009, www.smithsonian mag.com/air/the-political-history-of-cap-and-trade-34711212/#xzvAgUGF 4jwVpV2d.99.

24. Written testimony by Beth McGee, senior scientist at the Chesapeake Bay Foundation, to the United States Senate Subcommittee on Water and Wildlife, May 22, 2013, www.epw.senate.gov/public/_cache/files/8/1/81def80c-1a42-4411-adb 7-23a73a0aed05/01AFD79733D77F24A71FEF9DAFCCB056.52213hearingwit nesstestimonymcgee.pdf.

25. Ibid.

26. This quote is from the *Environment in Focus* radio program "Hogan Administration Proposes Pollution Credit Trading for Chesapeake Bay," November 18, 2015, http://wypr.org/post/hogan-administration-proposes-pollution-credit-trading -chesapeake-bay#stream/0.

27. Chesapeake Bay Foundation, "Nutrient Trading," www.cbf.org/issues/nu trient-trading.html?referrer=https://www.google.com/.

28. Theo Emery, "Fraud Case Shows Holes in Exchange of Fuel Credits," *New York Times*, July 4, 2012, www.nytimes.com/2012/07/05/us/biofuel-fraud-case -shows-weak-spots-in-energy-credit-program.html.

29. Ashley Seager, "European Taxpayers Lose €5bn in Carbon Trading Fraud," *Guardian*, December 14, 2009, www.theguardian.com/business/2009/dec/14/eu -carbon-trading-fraud.

30. US Environmental Protection Agency, "Pennsylvania's Trading and Offset Programs Review Observations, Final Report," February 17, 2012, www.epa.gov /sites/production/files/2015-07/documents/pafinalreport.pdf.

31. Pennsylvania Department of Environmental Protection, "Pennsylvania Chesapeake Watershed Implementation Plan: Phase 2," March 30, 2012, www.dep .state.pa.us/river/iwo/chesbay/docs/refmaterials/PAChesapeakeWIPPhase2_3 -30-12.pdf.

Accountability: The Bay Numbers Game

1. 1987 Chesapeake Bay Agreement, www.chesapeakebay.net/content/publica tions/cbp_12510.pdf.

2. Tom Horton, *Turning the Tide: Saving the Chesapeake Bay*, 2nd ed. (Washington, DC: Island Press, 2003).

3. Ibid.

4. There was also a 1983 bay agreement, but it just got the process started, so we won't spend much time discussing it. Chesapeake 2000 Agreement, www.chesa peakebay.net/channel_files/19193/chesapeake_2000.pdf.

5. Ibid.

6. Interview with Michael Wilberg, professor of biology at the Chesapeake Biological Laboratory of the University of Maryland Center for Environmental Science, April 24, 2014.

7. M. J. Wilberg, M. E. Livings, J. S. Barkman, B. T. Morris, and J. M. Robinson, "Overfishing, Disease, Habitat Loss, and Potential Extirpation of Oysters in Upper Chesapeake Bay," *Marine Ecology Progress Series* 436 (2011): 131–44, http://wilber glab.cbl.umces.edu/pubs/Wilberg%20et%20al%202011.pdf.

8. Maryland Department of Natural Resources press release, "Maryland's Oyster Population Continues to Improve, Highest since 1985," May 7, 2014, http://news.maryland.gov/dnr/2014/05/07/marylands-oyster-population-con tinues-to-improve-highest-since-1985/.

9. Chesapeake 2000 Agreement.

10. The Chesapeake Bay Program provides estimates on nitrogen, phosphorus, and sediment loads in the bay. These numbers are based on a combination of computer modeling and water quality monitoring stations in the bay. They are not corrected for rainfall or river flow. See Chesapeake Bay Program, "Nitrogen Loads and River Flow to the Chesapeake Bay," www.chesapeakebay.net/indicators/indicator /nitrogen_loads_and_river_flow_to_the_bay1.

11. Ibid. Note that these figures are not corrected for rainfall.

12. Ibid.

13. Ibid.

14. Ibid.

15. Chesapeake Bay Program, "Reducing Phosphorus Pollution," www.chesa peakebay.net/indicators/indicator/reducing_phosphorus_pollution. See also Chesapeake Bay Program, "Reducing Nitrogen Pollution," www.chesapeakebay.net/indi cators/indicator/reducing_nitrogen_pollution.

16. University of Maryland Center for Environmental Sciences EcoCheck Health Report Card on the Chesapeake Bay for 2015, http://ecoreportcard.org/ report-cards/chesapeake-bay/health/.

17. Chesapeake 2000 Agreement.

18. EPA Chesapeake Bay Program, "Underwater Bay Grass Abundance (Bay-wide)," www.chesapeakebay.net/indicators/indicator/bay_grass_abundance_baywide.

19. Virginia Institute of Marine Sciences, "Submerged Aquatic Vegetation (SAV) in Chesapeake Bay," http://web.vims.edu/bio/sav/AboutSAV.html.

20. Data on submerged aquatic vegetation come from the Virginia Institute of Marine Science/Robert Orth website on bay grasses, http://web.vims.edu/bio /sav/sav15/exec_summary.html. See also Chesapeake Bay Program, "Underwater Bay Grass Abundance (Baywide)."

21. Virginia Institute of Marine Science website on bay grasses, http://web.vims.edu/bio/sav/sav15/exec_summary.html.

22. EPA Chesapeake Bay Program, "Bay Barometer: A Health and Restoration Assessment of the Chesapeake Bay and Watershed in 2010," www.chesapeakebay.net/documents/cbp_59306.pdf.

23. Numbers from state data were obtained through a Maryland Public Information Act request to the Maryland Department of the Environment. This was also covered in the *Environment in Focus* radio program "The Illusion of Wetlands Restoration," March 13, 2013, http://programs.wypr.org/podcast/3-13-13-illusion-wetlands-restoration.

24. Numbers were obtained through inquiries to the US Army Corps of Engineers, Virginia Department of Environmental Quality, and Pennsylvania Department of Environmental Protection. See Tom Pelton, "The Illusion of Wetlands 'Restoration,'" *Bay Daily*, February 8, 2012, http://bit.ly/2cu3MjB.

25. David Moreno-Mateos, Mary E. Power, Francisco A. Comín, and Roxana Yockteng, "Structural and Functional Loss in Restored Wetland Ecosystems," *PLOS Biology*, January 24, 2012, http://journals.plos.org/plosbiology/article?id=10.1371/journal.pbio.1001247.

26. Maryland Department of the Environment, "Effectiveness of Maryland Compensatory Mitigation Program," December 2007, www.mde.state.md.us/programs/Water/WetlandsandWaterways/AboutWetlands/Documents/www.mde.state.md.us/assets/document/wetlandswaterways/Mit_Report_Title.pdf.

27. Chesapeake 2000 Agreement.

28. Data on forested buffers come from the US Forest Service. See Chesapeake Bay Program, "Planting Forest Buffers," www.chesapeakebay.net/indicators/indicator/planting_forest_buffers.

29. Ibid.

30. Chesapeake 2000 Agreement.

31. EPA Chesapeake Bay Program, "Bay Barometer."

32. Chesapeake Bay Program, "The State of the Chesapeake Bay: A Report to the Citizens of the Bay Region," June 2002, www.chesapeakebay.net/content/publications/cbp_13112.pdf.

33. Number calculated by examining the amounts transferred out of Program Open Space every year, according to the Maryland Department of Legislative Services annual "General Assembly 90 Day Reports," issued after each legislative session. The 2014 report, for example, can be found online at http://mgaleg.maryland.gov/Pubs/legislegal/2014rs-90-day-report.pdf.

34. Senator Thomas "Mac" Middleton, "Time to Lock Program Open Space," *Bay Journal*, March 2, 2016, www.bayjournal.com/article/time_to_lock_program_open_space.

35. US Government Accountability Office, "Chesapeake Bay Program: Improved Strategies Are Needed to Better Assess, Report and Manage Restoration Progress," October 2005, www.gao.gov/products/GAO-06-96.

36. David Fahrenthold, "Broken Promises on the Bay," *Washington Post*, December 27, 2008, www.washingtonpost.com/wp-dyn/content/article/2008/12/26/AR2008122601712.html.

37. September 16, 2016, interview with EPA Region 3 water protection division director Jon Capacasa for the *Environment in Focus* radio program "Defining Success in the Chesapeake Bay Cleanup," September 27, 2016, http://wypr.org/post/redefining-success-chesapeake-bay-cleanup.

38. David Fahrenthold, "Md. Watermen Mull Suing over Bay: Class Action May Focus on Farming Firms, Municipal Plants," *Washington Post*, July 28, 2004, www.washingtonpost.com/wp-dyn/articles/A20067-2004Jul28.html.

39. Interview with the Chesapeake Bay Program's associate director for science, Richard Batiuk, as part of an article I published in the *Baltimore Sun*; see Tom Pelton, "Standard to Measure 'Dead Zones' Changing: Redefining 'Low Oxygen' in Bay Raises Concerns," *Baltimore Sun*, May 12, 2005, http://articles.baltimoresun.com/2005-05-12/news/0505120032_1_batiuk-bay-program-oxygen.

40. Ibid.

41. University of Maryland Center for Environmental Science EcoCheck Chesapeake Summer Forecast, http://ian.umces.edu/ecocheck/forecast/chesapeake-bay/2015/.

42. Rebecca R. Murphy, W. Michael Kemp, and William P. Ball, "Long-Term Trends in Chesapeake Bay Seasonal Hypoxia, Stratification, and Nutrient Loading," *Estuaries and Coasts* 34 (2011): 1293, http://link.springer.com/article/10.1007/s12237-011-9413-7.

43. Chesapeake Bay TMDL, "Chesapeake Bay TMDL Executive Summary," www.epa.gov/sites/production/files/2014-12/documents/bay_tmdl_executive_summary_final_12.29.10_final_1.pdf.

44. Ibid.

45. Ibid.

46. September 16, 2016, interview with EPA Region 3 water protection division director Jon Capacasa for the *Environment in Focus* radio program "Defining Success in the Chesapeake Bay Cleanup."

47. Data come from EPA Chesapeake Bay Program model estimates for pollution loads entering the bay, corrected for rainfall and stream flow. Chart titled "Numbers from EPA Bay Program Model, Corrected 'Normalized' for Rainfall and Stream Flow," provided by EPA Region 3 spokesman David Sternberg via e-mail on September 16, 2016.

48. University of Maryland Center for Environmental Science EcoCheck Report Card on the Health of the Chesapeake Bay, http://ecoreportcard.org/report-cards/chesapeake-bay/health/.

49. Telephone interview with Jon Capacasa, EPA Region 3 water program director, September 16, 2016, conducted for the *Environment in Focus* radio program "Redefining Success in the Chesapeake Bay Cleanup," September 27, 2016, http://wypr.org/post/redefining-success-chesapeake-bay-cleanup.

50. Ibid.

51. Interview with Nicholas DiPasquale, director of the EPA Chesapeake Bay Program, November 1, 2016.

52. Interview with Richard Batiuk, associate director for science at the EPA Chesapeake Bay Program, November 1, 2016.

53. University of Maryland Center for Environmental Science EcoCheck Report Card on the Health of the Chesapeake Bay.

Conclusion: The Future of the Bay

1. Here I am referring to the banning of DDT in the United States, where mosquito-borne malaria was no longer a threat, not in Africa and Asia, where malaria is still common in some areas and where carefully controlled applications of the insecticide may still be needed for public health reasons.

2. Joshua Galperin, "'Desperate Environmentalism' Won't Save the Environment," *Los Angeles Times*, October 29, 2015, www.latimes.com/opinion/op-ed/la-oe-galperin-environmental-desperation-20151029-story.html.

3. US Geological Survey and partners, "New Insights: Science-Based Evidence of Water Quality Improvements, Challenges, and Opportunities in the Chesapeake," March 2014, www.chesapeakebay.net/channel_files/21409/new_insights_report.pdf.

4. Chesapeake Bay Foundation, "Debunking the 'Job Killer' Myth: How Pollution Limits Encourage Jobs in the Chesapeake Bay Region," December 2011, www.cbf.org/document-library/cbf-reports/Jobs-Report-120103-FINALe2ef.pdf. Environmental Integrity Project, "Don't Believe the 'Job Killer' Hype: Decades of Economic Research Show That Environmental Regulations Are Good for the Economy," January 16, 2017, www.environmentalintegrity.org/wp-content/uploads/2017/01/Jobs-and-environment-report.pdf.

5. US Bureau of Labor Statistics, "Extended Mass Layoffs in 2012," Report 1043, September 2013, table 4, www.bls.gov/opub/reports/mass-layoffs/archive/extended_mass_layoffs2012.pdf. These are "mass layoffs," defined as layoffs of 50 or more workers which are reported to the government by industry. Less than 0.2% of mass layoffs in 2010, 2011, and 2012 were caused by government intervention or regulations of any kind, including environmental regulations. Economist Eban Goodstein reports similar figures for the 1980s and 1990s in his book *The Trade Off-Myth: Fact and Fiction about Jobs and the Environment* (Washington, DC: Island Press, 1999), 1–7.

6. Chesapeake Bay Foundation, "Debunking the 'Job Killer' Myth."

7. Ibid.

8. Jerry Taylor, "Cheap Natural Gas, Not the EPA, Is Closing Old Plants," *National Review*, February 25, 2013, www.nationalreview.com/nrd/articles/340048/coal-meets-markets.

9. University of Maryland Center for Environmental Science EcoCheck Re-

port Card on the Health of the Chesapeake Bay, http://ecoreportcard.org/report
-cards/chesapeake-bay/health/. Interview with Bill Dennison, vice president for
science applications at the University of Maryland Center for Environmental Science, May 19, 2016.

10. Ibid. Readings of chlorophyll here are interpreted to indicate intensity of
algal blooms.

11. Maryland Department of Natural Resources press release, "Maryland's
Oyster Population Continues to Improve, Highest since 1985," May 7, 2014, http://
news.maryland.gov/dnr/2014/05/07/marylands-oyster-population-continues
-to-improve-highest-since-1985/. See also Maryland Department of Natural Resources, "Maryland Oyster Population Status Report 2015 Fall Survey," http://
dnr.maryland.gov/fisheries/Documents/FallSurvey-2015.pdf. In this discussion,
the oyster biomass index (on p. 18), an annual estimate of the total weight of oysters in an area, is interpreted as an approximate indicator of oyster populations.

12. Virginia Institute of Marine Science of the College of William & Mary, "SAV
in Chesapeake Bay and Coastal Bays," http://web.vims.edu/bio/sav/index.html.

13. University of Maryland Center for Environmental Science EcoCheck Report Card.

14. Many of these suggestions are not ideas I invented, but ones that have been
suggested by a variety of experts and lawmakers over the decades and compiled
here.

15. This is a concept that has long been advocated by Dr. Donald Boesch, president of the UMCES.

16. Karl Blankenship, "PA Plan Says It Will Increase Ag Inspections, Plant
More Trees," *Bay Journal*, February 28, 2016, www.bayjournal.com/article/pa_plan
_says_it_will_increase_ag_inspections_plant_more_trees. US Environmental Protection Agency Region 3 report, "Pennsylvania Animal Agriculture Program Assessment," February 2015, www.epa.gov/sites/production/files/2015-07/documents
/pennsylvania_animal_agriculture_program_assessment_final_2.pdf. Also, my April
6, 2016, *Environment in Focus* radio program "70 Percent of Pennsylvania Farmers
Violate Clean Water Law" included an interview with John Quigley, then secretary
of the Pennsylvania Department of Environmental Protection, who confirmed this
figure. See http://wypr.org/post/70-percent-pennsylvania-farmers-violate-clean
-water-law.

17. Rona Kobell, "Hogan Moves to Lift Rule for Less-Polluting Septic Systems
in MD," *Bay Journal*, September 18, 2016, www.bayjournal.com/article/hogan
_repeals_reg_requiring_less_polluting_septic_systems_in_md.

18. Interview with Nicholas DiPasquale, director of the Chesapeake Bay Program, November 1, 2016.

19. Karl Blankenship, "House Votes to Restrict EPA Oversight Power in Bay
Cleanup," *Bay Journal*, July 14, 2016, www.bayjournal.com/blog/post/house_votes
_to_restrict_epa_oversight_power_in_bay_cleanup.

20. Ibid. According to the *Bay Journal*, lawmakers from the Chesapeake Bay watershed voting for the amendment to prohibit the EPA from imposing punishments on states that fail to meet the pollution limits in the Chesapeake Bay TMDL were Representatives Lou Barletta (R-PA), Dave Brat (R-VA), Chris Collins (R-NY), Ryan Costello (R-PA), Charles Dent (R-PA), Bob Goodlatte (R-VA), Morgan Griffith (R-VA), Robert Hurt (R-VA), Evan Jenkins (R-WV), John Katko (R-NY), Tom Marino (R-PA), David McKinley (R-WV), Pat Meehan (R-PA), Alex Mooney (R-WV), Scott Perry (R-PA), Joseph Pitts (R-PA), Tom Reed (R-NY), Keith Rothfus (R-PA), Bill Shuster (R-PA), and Glenn Thompson (R-PA). Lawmakers from the Chesapeake Bay watershed voting against the amendment were Representatives Don Beyer (D-VA), John Carney (D-DE), Matthew Cartwright (D-PA), Barbara Comstock (R-VA), Gerry Connolly (D-VA), Elijah Cummings (D-MD), John Delaney (D-MD), Donna Edwards (D-MD), Randy Forbes (R-VA), Chris Gibson (R-NY), Richard Hannah (R-VA), Andy Harris (R-MD), Scott Rigell (R-VA), Dutch Ruppersberger (D-MD), John Sarbanes (D-MD), Robert Scott (D-MD), Chris Van Hollen (D-MD), and Robert Wittman (R-VA).

21. "Get rid of" EPA statement made by Donald Trump at March 3, 2016, Republican presidential primary debate hosted by Fox News in Detroit. See www .youtube.com/watch?v=GzPIbX1pzDg. Statement by Trump on eliminating "70 percent" of regulations reported in Chris Kaufman, "Republican Trump Says 70 Percent of Federal Regulations 'Can Go,'" *Reuters*, October 7, 2016, www.reuters .com/article/us-usa-election-trump-regulations-idUSKCN12629R.

Index